3rd Topical Conference on Heat Exchangers 2024

Topical Conference at the 2024 AIChE Spring Meeting and 20th Global Congress on Process Safety

New Orleans, Louisiana, USA
24-28 March 2024

ISBN: 979-8-3313-1717-1

Printed from e-media with permission by:

Curran Associates, Inc.
57 Morehouse Lane
Red Hook, NY 12571

Some format issues inherent in the e-media version may also appear in this print version.

Printed with permission by Curran Associates, Inc. (2025)

For permission requests, please contact AIChE
at the address below.

AIChE
120 Wall Street, FL 23
New York, NY 10005-4020

Phone: (800) 242-4363
Fax: (203) 775-5177

www.aiche.org

Additional copies of this publication are available from:

Curran Associates, Inc.
57 Morehouse Lane
Red Hook, NY 12571 USA
Phone: 845-758-0400
Fax: 845-758-2633
Email: curran@proceedings.com
Web: www.proceedings.com

TABLE OF CONTENTS

MODELING, SIMULATION, DESIGN, AND OPTIMIZATION OF HEAT EXCHANGERS

HEAT EXCHANGER FOULING MITIGATION

HEAT EXCHANGER CLEANING

NEW HEAT EXCHANGERS AND HEAT TRANSFER ENHANCEMENT TECHNOLOGIES

DIGITAL TRANSFORMATION, AI AND IOT TECHNOLOGY APPLICATIONS TO HEAT EXCHANGERS

COATING FOR HEAT TRANSFER APPLICATIONS

ELECTRICAL HEATERS

SPECIAL SESSION IN HONOR OF PROFESSOR KENNETH BELL

HEAT EXCHANGER CLEANING II

94a Experiences about the Trend Analysis of Critical Heat Exchangers within a Butadiene Unit

Marco Fazzini

The control of fouling in the extractive distillation section of the butadiene unit is a key aspect of this process. As a part of an optimization study conducted for a SABIC plant, a number of critical heat exchangers were assessed in terms of their thermal and hydraulic functioning. In particular, the solvent/solvent-hydrocarbon unit was rigorously modelled. Key indicators over various cycles were obtained using software for trend analysis. Such evaluations provided the base line for the validation of the cleaning effectiveness as well as the criteria for the rating of the plant at varying throughput. The presentation reviews the exchangers' simulation strategy and demonstrates the utilization of appropriate tools for analyzing the heat exchanger performance.

94b Optimization of Shell and Tube Heat Exchangers Operating in Heat Recovery Services with Tube Side Heat Transfer Enhancement

Peter Droegemueller, Hamzah Sheikh, William Osley

Optimization of heat exchangers for efficient heat recovery in thermal processes is an important aspect to reach net zero goals. To achieve high heat recoveries, a so-called temperature cross, where the cold fluid temperature intersects the hot fluid temperature, is encountered in these units. For shell & tube exchanger design, this results in a challenging set of conditions to work with.
Multi-pass shells in series configurations are common in situations with a small temperature approach, and in many cases are used in order to compensate for the temperature cross between the process streams. In such designs, the equipment and piping costs are high. Alternatively, a two pass F shell design is often considered, however here thermal and physical leakage across the longitudinal baffle and damaged sealing strips during bundle removal are major drawbacks. For these reasons F-type exchangers are frequently disallowed in end-user specifications. In general, the most cost-effective option is a single-pass E shell exchanger. However, the single pass E shell design is often not pursued, due to the following:

- Low tube side heat transfer due to low flow velocity in single pass design.
- Low tube side flow velocities often shift the flow regime from turbulent in multi-pass design into transitional or even laminar in single tube pass design.
- Operational issues in start-up and turn down conditions when operating near transitional flow regime.
- Bundle flow maldistribution often caused by very low frictional tube side pressure drop.
- Possibility of temperature stratifications (in-tube temperature pinch).

All these shortcomings can be overcome by using hiTRAN® Thermal Systems tube side heat transfer enhancement technology. Measured tube side heat transfer in laminar, transitional and turbulent flow are presented. The measurements in this region were conducted with different Grashof and Richardson numbers with and without tube side heat transfer enhancement. Results show that possible operational issues in the transitional flow region can be eliminated entirely. Over the entire operation range there is a linear response in terms of heat transfer with flow velocity (Reynolds number), as opposed to the response observed with empty tubes, thus eliminating the risk of sudden changes in the transitional flow region.

In addition, the much-improved tube side heat transfer of up to 16 times in laminar and 4 times in turbulent flow leads to smaller units with large capital cost savings.

The avoidance of possible in tube temperature stratification in low flow velocity single pass designs, when using hiTRAN® technology is explained in detail. In in-tube temperature pinch situations, large areas of effective heat transfer area is lost and leads to underperforming heat exchangers. This situation is difficult to predict and is not well captured with standard heat transfer correlations since it depends very much on complex mixed convection flow behaviour. In general, software design packages do not account for this situation and often require the user to make a judgement based on a simple raised warning message.
CFD flow videos in tandem with experimental measurements demonstrate the significant difference between plain empty tube and hiTRAN® enhanced flow in those situations. It is demonstrated that the tube side flow pattern is changed from stratified mixed convection flow to homogeneously forced convection flow thus greatly improving the predictability in terms of heat transfer and pressure drop.

As indicated, single pass S&T designs often lead to low flow velocities, with very low associated frictional pressure drop. This in turn can lead to uneven tube side flow distribution, especially in large bundles. Increased frictional pressure drop, generated by the use of hiTRAN® inserts, can lead to improved fluid distribution in the bundle due to the increased frictional pressure drop. By applying Computational Fluid Dynamics to real industry cases, the level of improvement is demonstrated.

The findings are rounded up by an industrial case study comparing multi-pass shells in series, two pass F shell and hiTRAN® enhanced single pass designs. The comparison calculations were performed using HTRI Xchanger Suite® and CALGAVIN's hiTRAN® plug in. In addition, HTRI Xchanger Optimizer® was used to evaluate reliably the cost implications of the different design.

Furthermore, in 2-phase heat transfer applications, hiTRAN® Thermal Systems is used to optimize industrial heat exchangers. This will be demonstrated with a comprehensive case study, where a large feed/effluent exchanger (Texas tower) is retrofitted on the tube side with hiTRAN® elements. The goal was two fold, an increased product throughput without the need to replace the heat exchanger and large energy savings with correspondingly large CO_2 reduction. Paraxylol is evaporated and superheated to around 327°C on the tube side before entering the fired heater for meeting reaction temperature in the reactor. After leaving the reactor the superheated vapour is then condensed on the shell side of the feed effluent exchanger transferring energy to the tube side.
A detailed analysis of the tube side flow and boiling regimes was undertaken using Aspen and HTRI design software. The results of the thermal and hydraulic results were then used to design a bespoke hiTRAN® insert geometry. In this case, two distinctive zones are present, the 2-phase region with good heat transfer and relatively high incremental pressure drop. Here an insert with low packing density was selected. In the subsequent superheated region, the heat transfer dropped (which is expected for gas flows) and the associated incremental pressure drop was lower. For that reason, a higher packed insert with higher enhancement characteristics were selected.

After in-situ installation during a planned refinery shut down, the measured performance increase was very much in-line with the predictions of the heat transfer and hydraulic model. The throughput was increased by around 18% and the heat recovery increased by around 3 MW. Since installation this equates to around 80,000 t of CO_2 savings.

94c Evaluation of Sloshing in Coil Wound Heat Exchangers Used in Floating LNG Applications

Bo Jin

Coil wound heat exchangers are used in most of the land-based, large-scale natural gas liquefaction facilities. They have a well-proven track record for performance and robustness. As the implementation of liquefaction processes extended to off-shore floating LNG facilities, it was important to investigate the effect of vessel motion on heat exchanger performance. One important aspect to consider was the potential for sloshing at locations in the exchanger where liquid is collected and distributed.

The flow characteristics are different from the classical study of sloshing where the fluid is enclosed inside a vessel (e.g. storage tank) with no flow in or out. The liquid in coil wound heat exchanger distributors continuously flow in and out of the system. To understand this key difference, a CFD model was developed to simulate the liquid motion inside a partially filled cylinder with and without continuous flow in and out. For a partially filled vertical cylinder without any continuous flow in and out, the CFD model was able to capture the well-known sloshing behavior documented in the literature. For the cylinder with a continuous steady in and out flow, the CFD simulation results demonstrate that the continuous flow suppresses sloshing. An analysis based on potential flow theory is described and confirms the CFD results, further demonstrating that flow in and out of the distributor changes the natural frequency of the system and thereby suppresses sloshing.

113a Condenser Design Performance Evaluation Under Dynamic Fouling Conditions with Hybrid-AI Digital Twins: An Industrial Case Study.

Himanshu Joshi, Emilio Diaz Bejarano, Francesco Coletti

Evaluating the expected performance of different heat exchanger designs once installed in the field can be challenging as a result of the wide variety of conditions they will encounter. Varying conditions include the inlet conditions of temperature, flowrate and pressure, but also the build-up of fouling over time, which will gradually degrade both the thermal and hydraulic performance of the equipment.

In the design and rating stages, the expected performance is normally assessed for selected cases, such as normal operation and turndown inlet conditions; and clean conditions and with standard fouling factors on both sides. Assessing other conditions is possible with standard design software, but becomes a quite tedious and time consuming activity.

In this context, Digital Twins capable of running dynamic simulations of process units represent a powerful alternative to screen a very wide range of conditions (both operational and fouling) to provide a map for the expected performance of heat exchangers. Not only that, once the Twin is set up, it can also be used to assess potential optimisation options, such as the use of bypasses, flow splits or cleaning scheduling. In cases where historical data is available for the service to study, it is even possible to evaluate the performance of different design using the actual plant conditions. In this way, the engineer can obtain all the information required to make the right decision and prepare for actual operation with much less effort.

In this work, an industrial case study is presented that focuses on a propane condenser in a refinery. Two options were considered to replace the old heat exchanger: 1) a new exchanger equipped with finned tubes, specially designed for this service but entailing relatively large capital investment; 2) a readily available heat exchanger with plain tubes which would require a minimum investment. A digital twin was setup for the case study and used to assess both options over a wide range of inlet conditions and gradual fouling build-up. The results showed that the less expensive option 2) could only be used for a narrow range of flowrate conditions and would, additionally, require frequent maintenance; option 1), on the other hand, would be expected to cope with the duty requirements over the entire range of expected conditions and even compensate for fouling build up over a substantial period of time without cleaning.

SPRING24 +20TH GCPS
A Joint AIChE and CCPS Meeting

Heat Exchanger Fouling in Renewable Diesel and Sustainable Aviation Fuel Production: Challenges & Solutions

Bryan Crom[1], Annapurna Sri Sowmya Turuga[1], Hitesh Bagaria[1], Xiomara Price[1], Delbert Grotewold[2], Anthony Fields[1]

[1] Veolia Water Technologies & Solutions, Tomball, TX
[2] Veolia Water Technologies & Solutions, Boulder, CO

Prepared for Presentation at
American Institute of Chemical Engineers
2024 Spring Meeting and 20th Global Congress on Process Safety
New Orleans, LA
March 24-28, 2024

Heat Exchanger Fouling in Renewable Diesel and Sustainable Aviation Fuel Production: Challenges & Solutions

Bryan Crom[1], Annapurna Sri Sowmya Turuga[1], Hitesh Bagaria[1], Xiomara Price[1], Delbert Grotewold[2], Anthony Fields[1]

[1] **Veolia Water Technologies & Solutions, Tomball, TX**
[2] **Veolia Water Technologies & Solutions, Boulder, CO**

Abstract

In the past few years, there has been a meteoric rise in Renewable Diesel and Sustainable Aviation Fuel production. This rise has been driven by economic incentives, subsidies, and national, state/provincial, and local governmental regulations requiring a reduction in carbon intensities of fuels. The emergence of these policies coincided with the conversion of a series of refinery units that had been idled due to petroleum-based fuel demand reductions caused by COVID-19 pandemic lockdowns and societal shifts to virtual ways of operation, resulting in the introduction of new economic value chains by transitioning these units to process renewable feedstocks. Typical feedstock sources for these renewable fuel units include: soybean and other vegetable oils, used cooking oils, distillers corn oil, recycled animal fats, and various blends of each of these components. Each of these sources have different market prices as well as carbon intensities associated with them, making each particular unit's feedstock blend of choice an optimization exercise similar to traditional petroleum crude blending optimization.

A typical renewable fuel processing unit may consist of a pretreatment section, a "renewable diesel unit" comprising a hydrotreating section, a water separation section, and an isomerization section, followed by a product recovery section. The industry has seen that these units are susceptible to many operational challenges, chief among them heat exchanger fouling. The significance of heat exchanger fouling in renewable fuel production is substantial. It exceeds mere operational disruptions, notably affecting the industry by increasing operational costs, diminishing efficiency, reducing unit run lengths, and posing potential risks to personnel and the environment. Effectively addressing and mitigating these challenges is crucial for cultivating safe, cost-effective, and environmentally conscious production methods.

This paper will provide a brief background on the rise of renewable feedstock processing units, present a primer on typical unit operations and feedstock types, and focus on a deep dive into some of the observed heat exchanger fouling types and the mechanisms that generated these foulants. Additionally, a discussion around Veolia's treatment solutions, including antifoulant chemistries, as well as operational experiences and expertise to counter these fouling issues will be presented.

Keywords: renewable diesel, heat exchanger fouling, fouling mechanisms, renewable fuels processing, sustainable aviation fuel, fouling mitigation, antifoulant

1 Introduction

The landscape of renewable diesel (RD) and sustainable aviation fuel (SAF) production is deeply intertwined with the dynamics of the global energy market, with the United States alone consuming a substantial 3.8 million barrels per day (bpd) of diesel and approximately 1.6 million bpd of jet fuel.[1] This underscores the vital role of these fuels in supporting transportation and industrial sectors. The surge in renewable energy production is driven by renewable energy credits and government incentives for greenhouse gas reduction, alongside regulatory mandates at various levels.[2]

Refineries have increasingly turned to fuels derived from renewable feedstocks, such as vegetable oils, animal fats and used cooking oil (UCO) to meet regulatory requirements and capitalize on market opportunities, leading to increased investment in renewable fuel production facilities. In the United States, vegetable oil and animal fat production amounts to approximately 420,000 bpd, with soybean oil alone contributing to around 240,000 bpd, a considerable portion of which is used for renewable fuel production, totaling about 110,000 bpd.[3] Alongside these feedstocks, more than 10 million tons of UCO is generated in the United States annually,[4] with the global market for UCO expected to reach $8.5 billion in the coming years.[5]

The composition of these feedstocks varies, with triglyceride molecules (composed of three long fatty acids' chains attached to a glycerol backbone) being prevalent in most cases,[6] while UCO may contain higher levels of free fatty acids and impurities due to its previous culinary use, posing unique challenges in the production processes.[7] Table 1 contains typical ranges for various properties of renewable feedstocks.[6] These properties serve as fundamental parameters influencing feedstock selection, process optimization and product quality. Regardless of the feedstock, the production of RD and SAF involves intricate processes that break down feedstock molecules into their constituent parts, yielding high-quality fuels.

Table 1. Typical Renewable Feedstock Properties.[6]

Property	Unit	Typical Range
Free Fatty Acids (FFAs)	%	2 - 80
Moisture	%	< 1.5
Insoluble Impurities	%	< 0.5
Unsaponifiables	%	< 1.5
Total Metals	ppm	5 – 275
Silicon	ppm	0 – 4
Iron	ppm	0.5 – 40
Potassium	ppm	1 – 25
Sodium	ppm	0 – 30
Calcium	ppm	1 – 90
Phosphorus	ppm	2 – 250
Chlorides	ppm	5 – 200
Sulfur	ppm	0 – 100
Nitrogen	ppm	0 - 700

Despite the promise of renewable fuels, challenges such as fouling and corrosion present significant hurdles throughout the production process. Understanding and mitigating these challenges is essential for ensuring the efficiency, reliability and sustainability of renewable fuel production. This paper focuses exclusively on fouling issues in renewable fuel processing, offering insights into fouling mechanisms and mitigation strategies.

2 Overview of Fouling Issues in Renewable Fuels Processing

The production of RD and SAF is not without its challenges, and fouling represents a significant issue throughout the processing stages. Fouling occurs due to the deposition of unwanted materials on heat transfer surfaces, leading to reduced efficiency, increased energy consumption and potential equipment damage. The understanding of fouling issues in petroleum refining is well-established, developed through extensive research and industry experience over several decades. This wealth of knowledge has enabled refineries to implement sophisticated fouling control measures and optimize process operations to minimize fouling-related downtime and maintenance costs. In contrast, the understanding of fouling in renewable fuels processing, such as RD and SAF production, is relatively emerging, with limited research conducted thus far. Understanding the mechanisms and factors contributing to fouling specific to the renewable fuels industry is essential for mitigating its impact on the production process.

Incorporating Veolia's capabilities in the renewable energy sector, this section draws both industry expertise and insights gained from literature on petroleum refining and process heat transfer principles,[8,9,10] to discuss potential fouling challenges at different stages of renewable fuel production from feedstock handling to final product distribution. While general literature on RD and SAF production offers valuable context,[11,12] additional insights were drawn from recent research, such as a study on fouling with lipids using hot wire testing and a thermal fouling test machine.[13]

2.1 Feedstock Preparation

In the initial stages of renewable fuel processing, feedstock preparation involves crucial steps such as milling and extraction. During milling, raw feedstock materials are ground to a fine consistency to increase the surface area of the feedstock facilitating subsequent extraction and conversion processes. In extraction processes, we aim to separate desirable components such as triglycerides and fatty acids from the feedstock matrix using solvents or mechanical methods. In both of these operations, fouling can occur due to the accumulation of organic residues and contaminants, which include fatty acids, triglycerides, proteins and particulate matter, on the internal equipment surfaces such as mill blades and extraction vessels. Additionally, factors such as moisture presence and temperature fluctuations further contribute to fouling by facilitating deposit formation and microbial growth.

2.2 Pretreatment

Pretreatment stage involves emulsification, water separation and other conditioning processes such as degumming and deacidification, to remove impurities from raw feedstock and enhance its suitability for further processing. Emulsification is primarily used to enhance the interaction between the oil phase and pretreatment agents. Feedstocks high in free fatty acids can lead to stable emulsions that are difficult to separate, causing fouling in downstream equipment such as centrifuges and settlers. Inadequate water separation from the oil phase after emulsification or other pretreatment steps, may allow water carryover into subsequent units, promoting deposit formation and corrosion. During degumming, phospholipids and other gums are removed from the feedstock through the addition of water and subsequent centrifugation or filtration. However, inadequate removal of these compounds can lead to fouling, particularly in heat exchangers and separators. Deacidification, which involves neutralizing free fatty acids in the feedstock with alkali agents, can also result in fouling due to the formation of soap deposits and other by-products.

2.3 Conversion

Following pretreatment, the feedstock undergoes catalytic reactions such as cracking, isomerization and Fischer-Tropsch synthesis, to produce hydrocarbons suitable for use as RD and SAF. Fouling can occur due to the deposition of reaction by-products and impurities such as coke, heavy polymeric species and sulfur compounds, resulting from thermal decomposition and polymerization reactions under high-temperature conditions, on reactor internals and catalyst surfaces. Additionally, nitrogen compounds present in renewable feedstocks can act as catalyst poisons or promote corrosion in certain process conditions. These phenomena lead to reduced heat transfer efficiency, increased pressure drop and decreased catalyst activity.

2.4 Refining and Blending

Fouling in renewable fuel refining and blending may not be as pronounced as in earlier processing stages. However, during these final stages of renewable fuel production, fouling can still occur due to the accumulation of persistent residual contaminants and reaction by-products in equipment such distillation columns, heat exchangers and storage tanks. In blending operations, the risks of fouling are primarily associated with mixing renewable fuels with conventional petroleum-based products. Incompatibilities between different fuel components can lead to phase separation or chemical reactions that result in fouling-related issues.

2.5 Utilities, Wastewater Management and Amine Sweetening

Within RD and SAF production facilities, the utilities, wastewater management and amine sweetening processes are susceptible to fouling from various sources, including organic and inorganic contaminants, microbial growth and chemical reactions. Fouling in utilities such as cooling water systems, influent water treatment units and steam systems can lead to reduced heat transfer efficiency, increased energy consumption and equipment corrosion. Additionally, fouling in wastewater treatment systems can hinder the removal of pollutants posing environmental risks and regulatory compliance challenges. Similarly, amine sweetening units, which are employed to remove acidic gases such as hydrogen sulfide from off-gas streams generated during various processing stages, are susceptible to fouling from degradation products, corrosion by-products and particulate matter which can compromise gas purification efficiency and increase operational costs.

3 Addressing Heat Exchanger Fouling in Renewable Diesel Unit: Case Study

In RD and SAF production, heat exchanger fouling is a multifaceted challenge arising from various factors, including the accumulation of organic contaminants and inorganic deposits. Understanding these complex mechanisms is crucial for devising effective mitigation strategies. To illustrate these challenges and implementation of practical solutions, the following case study presents how Veolia addressed heat exchanger fouling in a renewable diesel unit (RDU) through chemical treatment of a facility.

The exchanger showcased in Figure 1 (a) was part of the RDU processing a composite renewable feedstock. After three months of operation, Figure 1 (b) highlights significant fouling accumulation on the exchanger tube bundles.

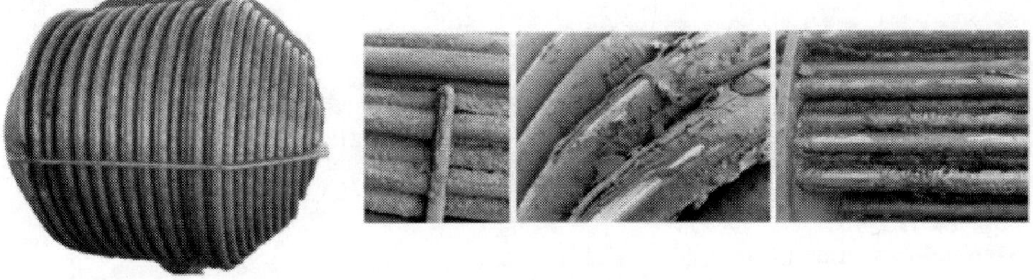

(a) (b)

Figure 1. Exchanger Tube Bundles in the RDU (a) Externally Clean (b) Fouling After 3 Months

3.1 Feedstock Characterization

The composite renewable feedstock processed through the exchanger comprised a blend of various feedstocks, including soy oil and distiller's corn oil (DCO), along with a recycle stream. Figure 2 illustrates the composite feed, showcasing its sticky, semi-fluid consistency and complex physical appearance at ambient conditions.[14]

Figure 2. Composite Renewable Feed Employed in this Study [14]

Veolia conducted a detailed characterization of the composite feed and its components used in the RDU process, presented in Table 2.[14] This characterization aimed to identify key parameters that could influence fouling in heat exchangers. The individual feedstocks (soy oil and DCO) which contribute to the composition of the composite feed, along with the recycle diesel stream reintroduced into the system from the production process, were also characterized to understand the contribution of each component to the process.

Table 2. Feedstock Characterization: Composite Feed and Individual Components [14]

Characteristic	Unit	Composite Feed	Soy Oil	Distiller's Corn Oil	Recycle
Total acid number	mg KOH/g	0.51	12.10	33.90	0.01
Basic nitrogen	wt%	< 0.001	0.002	0.007	0.002
Bromine number	g Br/100 g	8	43	29	< 0.1
Carbonyls	mg/L	68	0	213	6
Filterable Solids	ptb	10	54	18	4
Mercaptan Sulfur	wt%	< 0.003	0.011	< 0.003	< 0.003
Total Sulfur	wt%	0.007	0.247	0	0
Pyrrole Nitrogen	ppm	< 1	0	< 1	< 1
Aluminium	ppm	< 0.1	0.10	< 0.1	< 0.1
Calcium	ppm	< 0.2	0.70	< 0.2	0.2
Chromium	ppm	< 0.02	< 0.02	< 0.02	< 0.02
Copper	ppm	< 0.05	< 0.05	< 0.05	< 0.05
Iron	ppm	1.35	3.82	0.21	0.17
Potassium	ppm	< 0.5	< 0.5	< 0.5	< 0.5
Magnesium	ppm	< 0.1	0.20	< 0.1	< 0.1
Manganese	ppm	< 0.01	0.05	< 0.01	< 0.01
Sodium	ppm	< 0.5	1.00	< 0.5	< 0.5
Nickel	ppm	0.53	0.05	0.03	0.02
Lead	ppm	< 0.05	< 0.05	< 0.05	< 0.05
Vanadium	ppm	1.64	0.06	< 0.01	< 0.01
Zinc	ppm	0.2	1.17	0	0

Soy oil and DCO exhibited high total acid numbers (TAN), while the composite feed displayed a relatively low TAN value of 0.51 mg KOH/g, likely influenced by the contribution of the recycle stream which had a low TAN. However, the presence of carbonyls, with a concentration of 68 mg/L, and the moderate bromine number (8 g Br/100 g), suggested the potential for fouling due to the deposition of organic compounds and the likelihood of complex reactions such as polymerization and oxidation. The higher bromine number in soy oil compared to DCO, along with DCO's significantly high carbonyl content (three times that of the composite feed) highlights the individual contributions of these sources to the composite feed's fouling potential. Moreover, soy oil exhibited higher levels of filterable solids and metals, specifically iron, compared to other streams, indicating its potential for inorganic fouling. Although the filterable solids and metals in the composite feed were not exceptionally high, their presence, when combined with other factors, indicated a potential fouling risk likely due to their catalytic effect on oxidation and polymerization processes.

3.2 Deposit Analysis

The deposit analysis provides valuable insights into the composition of the fouling deposits accumulated in the heat exchanger. The predominant elements found in the deposit include carbon, hydrogen and oxygen as indicated in Table 3,[14] suggesting the presence of organic compounds and oxygen-containing functional groups. Additionally, sulfur, phosphorus, calcium, potassium, sodium, chlorine and other metals were present in smaller amounts, ranging from 0.1% to 4%. It's worth noting that sulfur can exist in organic as well as inorganic forms in renewable fuel production processes. The presence of these elements suggests a combination of organic contaminants and inorganic deposits contributing to fouling in the heat exchanger.

Table 3. Deposit Elemental Analysis [14]

Element or Metal	Wt %
Carbon	58.3
Hydrogen	7.5
Sulfur	4.0
Iron	4.0
Phosphorus	1.9
Calcium	1.1
Potassium	1.1
Sodium	0.7
Magnesium	0.6
Chlorine	0.1
Other Metals	0.8
Oxygen (by difference)	19.9

3.3 Fouling Control Testing with Veolia Treatment Program

The treatment testing phase aimed to assess the effectiveness of different dispersant and polymerization inhibitor chemistries, denoted as Products A, B and C, using a thermal fouling testing rig. This testing apparatus mimics the conditions of a heat exchanger by passing the sample through a shell-and-tube heat exchanger, where a heating rod serves as the heating element. The outlet fluid temperature was continuously monitored during the testing process to track changes over time. Fouling deposition occurs on the heating rod, leading to a decline in the outlet fluid temperature (ΔT) over time, indicating reduced heat transfer efficiency. The efficacy of each antifoulant product is determined by averaging temperature drops over a specified time period and expressing the percentage reduction in fouling relative to untreated conditions.

The fouling test provided valuable insights into the performance of Products A, B and C in mitigating fouling in heat exchanger systems. Despite temperature measurements being collected every minute, selective data points were labeled on the plot illustrating the variation in temperature differential (ΔT) over time (Figure 3) to enhance readability. Figure 3 reveals progressive fouling accumulation and reduced heat transfer efficiency in the untreated sample. All tested products contributed to reduction in fouling compared to the untreated sample. However, the extent of fouling mitigation varied among the treatments, with Product A consistently exhibiting the lowest increase in ΔT among the treated samples, as shown in Figure 3.

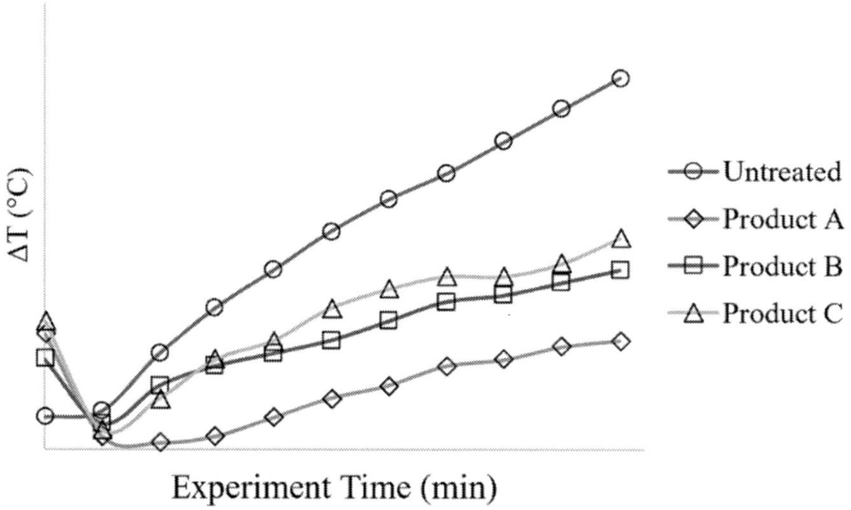

Figure 3. Temperature Difference (ΔT) vs. Time During Fouling Test [14]

Table 4 summarizes the percentage reduction in fouling for each antifoulant.[14] Product A contributed to an elevated reduction in fouling of 75% compared to the untreated sample, highlighting its superior performance in maintaining heat exchanger efficiency. In comparison, Products B and C exhibited moderate fouling reduction efficiencies of 49% and 42% respectively, relative to the untreated sample.

Table 4. Effect of Treatment on Fouling Reduction [14]

Treatment	% Fouling Reduction
Product A	75
Product B	49
Product C	42

3.4 Operational and Cost Benefits for Customer

Prior to implementing Veolia's treatment program, the producer's operations were significantly disrupted by fouling from the polymerization of feedstock on the feed side of the feed-effluent heat exchanger. This led to frequent unit shutdowns for cleaning, varying between 34 and 81 days as depicted in Figure 4.[14] On average, the shutdowns occurred approximately every 60 days for cleaning, translating to an average of six shutdowns per year. Each shutdown resulted in 3 days of lost operation, leading to significant financial losses for the facility estimated at $6-8 million per shutdown.[14]

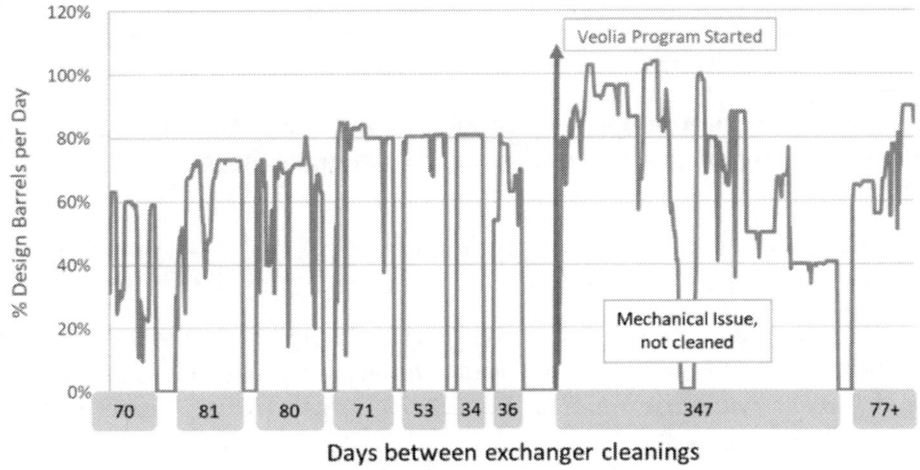

Figure 4. Comparison of Exchanger Cleaning Frequency Before and After Treatment Program Implementation [14]

Following the implementation of Veolia's treatment program with Product A, notable improvements were observed in heat exchanger performance. The RDU operated for over 340 days without any shutdown associated with exchanger fouling. Although there was a shutdown during this period, as shown in Figure 4, it was not due to the performance of the treatment program but was caused by an unrelated mechanical issue or maintenance activity. The treatment avoided four shutdowns per year over a 12-month period, resulting in additional revenue to the producer of $22-30 million.[14] Furthermore, the decreased fouling reduced the need for cleaning, resulting in a decrease in cleaning frequency and annual cleaning costs, considering a cost of $40,000 per cleaning.

4 Conclusions

This study sheds light on the critical challenges associated with fouling in renewable fuels processing and presents key findings obtained from an in-depth case study. Through meticulous characterization of feedstocks, deposit analysis and fouling control testing with Veolia's treatment programs, valuable insights into potential fouling mechanisms and their impact on heat exchanger performance have been gained.

The findings from this work emphasize the importance of proactive fouling management to maintain operational efficiency and minimize downtime in renewable fuels processing facilities. By recommending tailored solutions for fouling mitigation, substantial cost reduction and operational benefits for producers can be realized.

As the renewable energy sector continues to evolve, vigilance in overcoming fouling challenges will be crucial to unlocking the full potential of sustainable energy sources. Looking ahead, ongoing research and innovation will continue to play a pivotal role in developing more effective strategies for fouling mitigation and optimizing renewable fuels processing operations.

5 References

[1] U.S. Energy Information Administration. February 2024 Monthly Energy Review. http://www.eia.gov/totalenergy/data/browser/. Accessed March 9, 2024.

[2] U.S. Energy Information Administration. Renewable Energy Explained. http://www.eia.gov/energyexplained/renewable-sources/incentives.php. Accessed March 9, 2024.

[3] U.S. Department of Agriculture. U.S. Bioenergy Statistics. http://www.ers.usda.gov/data-products/u-s-bioenergy-statistics/. Accessed March 9, 2024.

[4] Chong JW, Chemmangattuvalappil NG, Thangalazhy-Gopakumar S. Aviation Biofuels: Conversion Routes and Challenges. *Sustainable Technologies for the Oil Palm Industry: Latest Advances and Case Studies.* Foo DCY, Aziz MKTA, Yusuf S. Springer Nature, Singapore 2023; pp. 33–85.

[5] WH, Koay SS, Chia SR, Chia WY, Tang DY, Nomanbhay S, Chew KW. Recent Advances in the Conversion of Waste Cooking Oil into Value-Added Products: A Review. *Fuel* 2022; **324**: 124539.

[6] Sulzer Chemtech. *Pretreating Renewable Feeds to Produce Sustainable Clean Fuels.* http://www.sulzer.com/-/media/images/products/gtc/pretreatment_whitepaper.pdf. Published 2022. Accessed March 9, 2024.

[7] Di Pietro ME, Mannu A, Mele A. NMR Determination of Free Fatty Acids in Vegetable Oils. *Processes* 2020; **8(4)**: 410, DOI: http://doi.org/10.3390/pr8040410.

[8] Speight JG. *Fouling in Refineries.* Gulf Professional Publishing; 2015.

[9] Kondyli A, Schrader W. Understanding "Fouling" in Extremely Complex Petroleum Mixtures. *ACS Appl. Energy Mater.* 2020; **3(8)**: 7251–7256.

[10] Serth RW. *Process Heat Transfer: Principles and Applications*. Elsevier Academic Press; 2007.

[11] Kumar D, Singh B, Gupta SK. *Renewable Diesel: Value Chain, Sustainability, and Challenges*. Elsevier; 2023.

[12] Ng KS, Farooq D, Yang A. Global Biorenewable Development Strategies for Sustainable Aviation Fuel Production. *Renewable Sustainable Energy Rev*. 2021; **150**: 111502.

[13] Byman O. Fouling Caused by Fats and Oils: The Suitability of Falex 450 Thermal Fouler for Fats and Oils. Master's thesis. Turku University of Applied Sciences, 2020. Accessed March 9, 2024. http://www.theseus.fi/bitstream/handle/10024/338969/Byman_Olli.pdf.

[14] Grotewold D, Wells K. Best Practices to Face Challenges of Processing Renewable Diesel. *AFPM Summit*. October 18-20, 2022; San Antonio, TX.

113c Progress Towards a SAFT Based Crude Oil Fouling Model

Aaron Smith

A major mechanism of crude oil fouling is asphaltene precipitation. Numerous threshold fouling models have been proposed that require regression to measured data to be applied. These models have utility when working with historical data and users wish to fit models to extrapolate trends into the future for optimization of cleaning and maintenance. By their empirical nature, these models do not explain the underlying mechanism based on the chemistry and thermodynamics of the crude oil constituents. To gain a more fundamental understanding of asphaltene precipitation, HTRI and ENNOVA collaborated to develop a Statistical Associated Fluid Theory (SAFT) model for crude oil that can predict the amount of asphaltenes that precipitate for a given temperature and pressure. The SAFT model is the basis for a crude oil fouling model that assumes the rate of asphaltene precipitation to form primary particles (100-200 nm) is proportional to the degree of super saturation of asphaltenes. primary particles are assumed to cause fouling. primary particles can continue to grow to form larger aggregates that are assumed to not foul. This approach has been successfully applied in upstream context. This theory also explains trends in HTRI fouling data. An overview of the theory, characterization methods, modelling approach and results to date will be provided.

182a Analysing the Influence of Macro-Economic Factors on Cleaning Decision-Making: A Case Study in the Petrochemical Industry

Emilio Diaz Bejarano, Hans Borgt, Francesco Coletti

Analysing the Influence of Macro-economic Factors on Cleaning Decision-Making: A Case Study in the Petrochemical Industry

182b How Contractors CAN Leverage Automation to Introduce Outcome-Based Pricing

Kelley Crush

Smart, data-producing automated solutions are revolutionizing the industry, offering many benefits to businesses and end customers. Increasingly, customers seek partnerships with contractors that are investing in technology and automation to reduce downtime, gain efficiency, and prioritize safety. Time and material-based pricing models focus on activity – not productivity. Outcome-based pricing creates win-win results for contractors and end customers who are looking for better productivity and return on investment. This presentation will explore how contractors can leverage data and automation to introduce outcome-based pricing to their end customers.

134a Tubular Enhancement for Laminar Flow Reactors Using Static in-Line Devices

Hamzah Sheikh, Peter Droegemueller

Introduction

Increasing efficiency on industrial plants is an ongoing endeavour, with efforts to reduce wastage, carbon emissions and production times. As a result, more and more plants are opting to use tubular reactors in place of traditional batch reactors, moving to more continuous processing.

The difference in flow mechanics results in a change in process requirements, and if not considered carefully can lead to poorly designed units either too large or underperforming. A difficulty in processing using tubular flow for reactions can be a result of inhomogeneity, where reactants are not well dispersed in their solvent. Particularly in high viscosity fluids, a high axial dispersion can be observed resulting in a non-ideal profile dissimilar to plug flow. Furthermore, poor heat transfer can result in non-uniform reactions, leading to poor product yields.

In-line static mixers and turbulators have been used to improve these issues, particularly where flow is viscous and prone to poor heat transfer and mixing. A range of devices are commercially available, predominantly static mixers, which enhance exchanger performance at the cost of pressure drop. These devices are often designed to cut the flow and incorporate sections of fluid into one another through forced channels or paths, and although these inserts can improve homogeneity of the fluid, a significant amount of energy is lost through friction and therefore requires more pumping power. Furthermore, improving mixedness does not necessarily result in better heat transfer or residence time.

We investigated the advantages of using two types of in-line turbulator in viscous applications, a wire matrix-based insert (hiTRAN), and a new plate type insert (hiVISC). These devices are designed for enhancing performance in tubular reactors, specifically improving residence time distributions and tube-side heat transfer.

Methodology

For this study, a combination of experimental and theoretical techniques were utilised. These methods measured the effect of tubular devices on heat transfer, pressure drop, residence time and gave an indication of mixing.

Heat transfer and pressure drop measurements

Heat transfer enhancement and pressure drop penalty was measured using the high viscosity flow rig (HVFR), consisting of a heating test section and a cooling test section which were operated simultaneously. Each test section consisted of a tube-in-tube configuration, with water acting as the service fluid on the shell side flowing counter currently. Glycerol was used as the process fluid due to its Newtonian behaviour and range of viscosities at the operational temperatures between 10 - 60 ^{o}C. Measurements of tube side and shell side temperature were taken on both inlets and outlets. Each test section was also fitted with a differential pressure sensor, which measured the in-tube pressure drop. The HVFR was also fitted with a Coriolis flowmeter and a bypass section to measure glycerol viscosity.

Experiments were conducted by inserting a tubular enhancement device into each test section and measuring the heat transfer over a range of flowrates and temperatures by adjusting the gear pump frequency using a frequency inverter. Temperatures were controlled using two separate flow loops, one for cooling coupled with a chiller unit and one for heating which used a controlled tank with an immersion heater.

Verified computational fluid dynamics (CFD) models were used to extend the scope of study for each of the devices. A computational model of the HVFR test sections were constructed in Ansys Fluent, and extensively calculated to minimise errors. The computational setup mimicked the conditions of HVFR tests, where heat transfer and pressure drop results accurately matched those observed in experiments. These results were then extended into deep laminar flow, by computationally increasing the fluid viscosity and thereby reducing the Reynolds number by an order of magnitude.

Mixing measurements

Mixedness was measured using planar laser induced fluorescence (PLIF), where a primary fluid is mixed with a secondary fluid doped with a fluorescent material and a plane is imaged using a laser sheet and camera. As the secondary fluid fluoresces under laser light, the resulting image consists of bright and dark areas representing each of the species which can then be computationally analysed to derive the mixing effectiveness of the enhancement device being tested.

The planar laser induced fluorescence rig (PLIFR) used glycerol as both the primary and secondary fluids, where the secondary fluid was doped with Rhodamine for the fluorescent effect. Flow was temperature controlled using a heating/cooling unit and a plate exchanger, giving a range of viscosities for glycerol and thereby a range of Reynolds numbers ranging from 0.03 - 16.5 (7.5 - 55 ^{o}C).

Pure glycerol flow was established first, followed by a central flow injection prior to the 23 mm internal diameter by 300 mm long test section. Imaging occurred after the testing section, where a laser sheet illuminated a plane through a clear Perspex section. Pressure drop across the tube internals was measured using a differential pressure sensor.

The resulting images had 13 μm spatial resolution, with 4096 levels of brightness. 50 images were captured for each test, combined, calibrated, and normalised to give a final result. Concentration-brightness linearity and temperature independence tests indicated pixel brightness was only affected by concentration of the secondary fluid.

Results and conclusions

Pressure drop

Baseline experiments were conducted using the HVFR with no tube internals, where isothermal results in pressure drop matched theoretical expectations. Experiments were conducted for $Re = 0.1$ - 100, by adjusting the pump frequency and changing the temperature of glycerol. Inserts increased the pressure drop observed in the test sections. For hiTRAN, increasing the packing density resulting in an increase in pressure drop, and the same trend was observed when reducing the pitch for hiVISC.

Heat transfer

As with pressure drop tests, initial baseline experiments for tube side heat transfer were conducted for the test sections without internals using the HVFR. The same tests were then conducted for each of the inserts. All inserts increased the heat transfer coefficient in comparison to the empty tube. Low density hiTRAN had the lowest heat transfer enhancement, which increased with packing density. Similarly, heat transfer coefficient increased with a reduction in pitch for hiVISC.

Heat transfer efficiency

An efficiency for each insert was calculated by computing a ratio between heat transfer enhancement over pressure loss in comparison to the empty tube results. Heat transfer coefficient was made dimensionless by calculating j-factor:

$$j = Nu\ Pr^{-n}$$

Dimensionless pressure drop was calculated using Darcy friction factor, *f*:

$$\Delta P / L = (f \rho v^2) / (2 D)$$

Finally, insert efficiency, η, was calculated as follows:

$$\eta = [(j_{insert} / j_{empty}) / (f_{insert} / f_{empty})^{m}]$$

Using this definition, insert performance could be evaluated against the empty tube. $\eta = 1$ represented empty tube performance, whereas an increased value indicated increased heat transfer with less pressure penalty than an equivalent length of empty tube required for the same heat transfer.

hiTRAN inserts show an increasing η with increasing *Re*, indicating best performance where $Re > 50$, particularly for high packing density. hiVISC shows up to $\eta = 2.3$ at very low *Re* (< 1), which increases as *Re* increases. Trends show η may reduce as $Re > 100$. Both sets of inserts show advantages over commonly used helical type inserts often used for viscous heat transfer enhancement.

Mixing efficiency

As with heat transfer efficiency, an efficiency for mixing can be calculated to compare insert performance. By computing the coefficient of variance (*CoV*) for images obtained using the PLIFR, an indication of mixedness can be evaluated against friction factor *f*. Mixing efficiency, λ, can be defined as:

$$\lambda = CoV f^{-1/3}$$

Tested inserts resulted in high values for λ, indicating an efficient mixing capability. This, however, corresponded to a high overall *CoV*, i.e. relatively poor mixing in comparison to other commercially available static mixers. Both hiTRAN and hiVISC act as low-intensity mixers, for which the intensity can be adjusted by varying packing density and pitch, respectively.

Conclusion

Using a variety of experimental methods, we characterised two types of tubular enhancement device (hiTRAN and hiVISC) in terms of pressure drop, heat transfer enhancement and mixing capability for high viscosity applications. Results showed both sets of inserts could deliver efficient enhancement, where pressure drop penalty was a fraction of the heat transfer enhancement. Applying these devices in flow reactors, viscous heat exchangers and other units can provide significant benefits in reducing costs, footprint, energy usage and carbon emissions.

134c How Can We Design Innovative and Efficient Heat Exchangers Via Additive Manufacturing?

Simone Mancin, Giacomo Favero, Adriano Pepato

Recently, Additive Manufacturing (AM) of metal components has opened new frontiers in heat transfer applications in multiple fields including chemical processes, going beyond the capabilities of conventional technologies. Despite the great design freedom offered by AM, when dealing with metals, there are a few issues that should be considered to exploit the great capabilities of this manufacturing technology: design for additive, surface roughness, thermo-physical properties of printed components, etc.

This presentation discusses, in detail, how to design the heat exchangers to be manufactured by 3D metal printing for real applications and not just to become fancy paperweights.

134d Enhanced Heat Transfer for Shell and Tube Heat Exchangers in the Process Industry – the New Standard

Thomas Lang, Lotfi Redjem Saad

Enhanced heat transfer technologies for shell and tube heat exchangers (STHX) are well established for decades and bring value in many industries such as the air conditioning, refrigeration and heating as well as the hydrocarbon processing and power industry.

The technologies referred to are the well-known low-fin type tube with a varying fin density on the tube outside as well as dual enhanced tubes. The low-fin tube technology is thereby responding and complying with the need to optimize the limited shellside heat transfer for a whole variety of heat transfer functions such as 1-phase liquid and gas as well as boiling and condensing, however restricted to optimize the heat transfer on the shellside only.

This limitation and need for the optimization of the overall heat transfer is addressed by a new generation of dual enhanced tubes with the combination of a low-fin resp. dedicated boiling and condensing structures on the outside and an enhancement structure on the inside of the tube.

In the last two to three decades a large variety of these heat transfer technologies has been established as "new standard" both for brown and greenfield applications together with a fully established industry reference basis. Key features are the application of industry relevant tube materials, compliance with code and industry requirements, establishment of solutions for manufacturing and operational aspects as well as a fully qualified and proven supply chain.

In a few selected EPC projects, the benefits in using superior heat transfer technology for improved CAPEX and OPEX and along plot space reduction, energy saving and $CO2$ footprint reduction as well as production improvements are shown.

169a A Machine-Learning-Based Framework for Data Regression of Heat Transfer Data

Jose Loyola-Fuentes, Nima Nazemzadeh, Emilio Diaz Bejarano, Simone Mancin, Francesco Coletti

Machine Learning (ML) algorithms are emerging in various industries as a powerful alternative to traditional data regression methods. A major reason is that, unlike deterministic models, they can be used even in the absence of detailed phenomenological knowledge. Not surprisingly, the use of ML algorithms is being explored also in heat transfer applications. It is of particular interest in systems dealing with complex geometries and underlying phenomena (*e.g.* fluid phase change, multi-phase flow, heavy fouling build-up). However, heat transfer systems present specific challenges that need addressing, such as the scarcity of high-quality data, the inconsistencies across published data sources, the complex (and often correlated) influence of inputs, the split of data between training and testing sets, and the limited extrapolation capabilities to unseen conditions. In an attempt to help overcome some of these challenges and, more importantly, to provide a systematic approach, this presentation proposes a regression framework for the deployment of ML-based regression to estimate key quantities (*e.g.* heat transfer coefficient), to be used for improved design and operation of heat exchangers. The framework consists of six steps: i) data pre-treatment, ii) feature selection, iii) data splitting philosophy, iv) training and testing, v) tuning of hyperparameters, and vi) performance assessment with specific indicators, to support the choice of accurate and robust models. A relevant case study involving the estimation of the condensation heat transfer coefficient in microfin tubes is used to illustrate the proposed framework. Two data-driven algorithms, Deep Neural Networks and Random Forest, are tested and compared in terms of their estimation and extrapolation capabilities. The results show that ML algorithms are generally more accurate in predicting the heat transfer coefficient than a well-known semi-empirical correlation proposed in past studies. In addition, feature selection enables simpler models that depend only on features that are potentially most related to the target variable. Special attention is needed however, as overfitting and limited extrapolation capabilities are common difficulties that are encountered when deploying these models.

169b Linking CFD Simulation to Continuum Models Using Correlation Equations for Mass and Heat Transfer in Chemical Engineering

Andraz Pavlisic, Matej Huš, Blaž Likozar

The reaction mechanism and the surface properties are still not fully understood, despite the technological advances. Therefore, the exact process of the reaction and the interaction of the catalytic surface have been a focus of academic research in recent years. A multi-scale hierarchical approach has been successfully applied to a complex industrial process, where a heterogeneous catalytic system is modelled from atomistic to macro level. However, this approach is unlikely to become a mainstream method for solving industrial processes, because it requires a lot of computational resources.

Computational fluid dynamics (CFD) is a powerful tool that can be used to tackle problems in chemical engineering, especially those involving fluid flow, heat and mass transfer, and chemical reactions. CFD can provide detailed information about the local phenomena that occur in complex systems, such as temperature, pressure, velocity, concentration, and reaction rate. CFD can also help to design and optimize chemical processes and equipment, such as reactors, separators, mixers, and heat exchangers. One of the challenges of using CFD in chemical engineering is to link the results of CFD simulation to the continuum models that are commonly used in engineering practice. These models assume of homogeneous properties and behaviour of the fluid, and they use correlation equations (CE) to describe the mass and heat transfer coefficients. CE are empirical formulas that relate dimensionless numbers and coefficients to predict the mass and heat transfer rates in a system. CE are useful because they simplify the complex three-dimensional phenomena to one dimension, and they can be used in much easier models, such as plug flow reactor (PFR) or continuous stirred tank reactor (CSTR).

However, CE have some limitations and drawbacks. First, they have to be determined experimentally for each specific system, which can be very expensive and time-consuming, especially for industrial scale reactors. Second, they are only valid for a certain range of the operating conditions of the system, such as Reynolds number, Prandtl number, Schmidt number, etc. Third, they do not capture the local variations and fluctuations of the properties and behaviour of the fluid, which can be important for some systems.

Therefore, a novel technique has been proposed to link CFD simulation to continuum models using CE. The idea is to use CFD simulation to calculate the local phenomena that are needed to determine CE, such as Nusselt number, Sherwood number, Stanton number, etc. Then, these values are averaged over the cross-section or volume of the system to obtain the effective CE that can be used in simple continuum models. This way, CFD simulation can provide more accurate and reliable CE that can account for the effects of geometry, flow pattern, turbulence, reaction kinetics, etc. Moreover, this technique can reduce the need for experimental data and extend the validity range of CE. With this technique, CFD simulation can be linked to pseudo-homogeneous and heterogeneous continuum models that can describe the mass and heat transfer rates in a system. This can enable fast and targeted optimization of industrial processes by changing the operating conditions or design parameters of the system. Some examples of industrial processes that could benefit from this technique are gas-liquid reactors, spray dryers, nanofluid heat exchangers, etc.

150a Novel Experimental Setup for Fouling Measurements and Evaluation of Mitigation Technologies

Simone Mancin, Francesco Coletti

Development of a rig for fouling measurements and testing of mitigation technologies

150b Multifunctional Coatings to Eliminate Exchanger Fouling and Improve Refinery Output

Katherine Uttley, Anika Galvan, Alexander Ventura, Matthew Nakatsuka, Jake McHenry

Fouling management of heat exchangers used in petroleum refining is a source of significant economic loss and carbon emissions. The high level of contaminants contained within crude oil (salt, wax, sand, acid, mud, corrosion products) means that the deposition mechanics can be quite complex, and fouling can be quite rapid, leading to thermal efficiency losses of up to 30% (Nakao, 2017, Costa, 2011). An additional consideration is the wide temperature range of the pre-heat train (PHT), where during the first main step of the refining process, crude oil is heated from transport temperatures of approximately 30°C to temperatures exceeding 400°C, excluding the use of many typical polymeric anti-fouling polymeric materials.

This talk will cover three different sets of "omniphobic", fouling-resistant surface treatments optimized for different stages of the PHT. These materials all utilize a similar low-surface energy chemistry combined with extremely low finished surface roughness to provide broad repulsion behavior against all the different particulates within the crude oil product. The three materials described here vary predominantly in their compatible temperature ranges: the first, designed primarily for temperatures up to 225°C, has been fully deployed in operational systems, and has demonstrated complete repulsion of adhered fouling for greater than 1 year. The second material, designed for temperatures up to 375°C, has been applied to subscale shell-and-tube heat exchangers on a trial basis on both the interior and exterior. The final material, currently under benchtop evaluation, has shown material durability and adhesion comparable to existing commercially available coatings (following ASTM D3359 and ASTM D4060, respectively), and temperature compatibility up to 550°C.
Simulated exposure testing on all three materials showed compatibility with a broad gamut of typical solvents used in cleaning operations. Utilization of such coatings has suggested the overall lifetime of PHT systems could be increased 2-fold, lowering costs substantially without capital investment or retooling of existing processes.

150c SMART and Multifunctional Tubacoat Technology for OPEX, CAPEX Reduction, Decarbonization and Biofuels Production in Refining and Petrochemical Applications.

Sanjay Lodha

TUBACOAT is a unique disruptive proven technology for refining, chemical and petrochemical critical applications processing difficult feedstock including biomass, a game changer for the industry for OPEX and CAPEX reduction, decarbonization and biomass processing. This Innovative and customized silica-based inert coating technology from Tubacex Group provides anti-fouling, anti-corrosion and anti-coking properties in harsh environments and extreme temperatures (800°C/1472°F), making TUBACOAT useful in almost all refineries and petrochemical applications, focusing on delayed cokers, visbreakers, cracking, crude vacuum distillation unit furnaces, resid processing, biomass processing, pyrolysis and similar units.

In Most refining conversion units such as delayed coker and others, Unscheduled shutdowns, high maintenance, reduced heat transfer, higher fuel consumption and throughput reduction due to furnace and heat exchanger tube coking, fouling and corrosion is a common problem. TUBACOAT technology minimizes formation, deposition and plugging due to coke/carbon in addition to reduction in fouling, corrosion and fuel usage. Use of TUBACOAT technology in these application leads to longer runs, incremental life cycle for equipment, reliable operations, increased heat transfer, reduced fuel consumptions, reduced CO_2 emissions and hence significant savings in OPEX and CAPEX.

Results and case studies from various commercial field trials and runs with major US, Europe, Asia refineries and new applications using Tubacoat technology covered in this presentation. Results include advantages of resistance of ceramic coated tubes to various forms of coking, corrosion, fouling, scaling, plugging, abrasion and the expected cost savings.

73a Gas-to-Electric Heating System Conversions for Decarbonization

Art Moslow

As the world strives to limit global warming and reduce its carbon footprint, the electrification of industrial heating processes is a critical step towards a more sustainable future.

The transition from a gas to an electric heating system is facilitated by a wide, unique range of metallic and ceramic materials (including NiCr, FeCrAl, Sic and MoSi2). These materials cover a broad spectrum of temperatures and atmospheres. Their careful selection and design allow for tailored heating solutions that meet specific demands within the Steel, Petrochemical, Glass and Heat-Treat industries, to name a few.

These industrial sectors face challenges from carbon emissions, volatile energy prices and growing energy demands all while being formidable contributors to the global carbon footprint. Some electrified heating solutions for decarbonization have successfully been implemented in continuous annealing and galvanizing lines, pit furnaces for ingot heating, walking beam furnaces for billet preheating, and roller hearth furnaces for bearing heat-treating. These electric heating systems significantly reduced CO2 emissions and offered operational benefits such as improvements in efficiency, temperature and working environments.

73b Enhanced Baffle Design for Electric Heat Exchangers in Gas Applications

Dave Johnston

White Paper and Presentation Abstract for 2024 AIChE Spring Meeting Topical Conferences

Topical 11 – Third Topical Conference on Heat Exchangers

Enhanced Baffle Design for Electric Heat Exchangers in Gas Applications

Thermon is pleased to present its innovative implementation of Optimized Baffle Solutions for Electric Heat Exchangers in Gas Applications at the upcoming 2024 AIChE Spring Meeting Topical Conferences. This presentation highlights the development, testing, and validation of our unique baffle designs, which offer superior heat transfer performance, extended equipment life expectancy, and significant cost reductions.

The discussion focuses on the comprehensive testing and validation process employed to ensure the effectiveness and reliability of our new heater designs equipped with innovative flow modifiers. Through a combination of advanced software simulations using HTRI and Ansys CFD, as well as empirical test data obtained from our dedicated R&D test set-up, we have successfully verified the performance and efficiency of the optimized baffle design.

Our findings demonstrate remarkable enhancements in heat transfer coefficients, resulting in up to 4 times improved heat transfer efficiency compared to conventional designs. The optimized baffle design also enables higher heat fluxes, further maximizing heat transfer capacity and efficiency without compromising heater longevity, shell design temperatures and process fluid limitations.

Beyond performance improvements, our research has shown that the new baffle design significantly extends the life expectancy of electric heat exchangers, ensuring long-term reliability and reduced maintenance costs. Furthermore, with up to a 60% reduction in overall costs, the implementation of our optimized baffle design offers substantial economic benefits to end-users.

The presentation will delve into the detailed testing procedures, validation methodologies, and performance comparisons with conventional designs. Key aspects such as design considerations, installation guidelines, and operational implications will also be discussed, providing attendees with practical insights for applying this innovative solution in their own gas heat exchange applications.

Keywords: heat transfer, higher heat flux, longer life expectancy, cost reduction, improved heat transfer coefficients, electric heat exchangers, gas applications, baffle placement, validation process.

73c Decarbonization of Fossil Fuel Process Heaters through Heating Systems Deploying Electric Heat Exchangers with Closed Loop Process Control

Dennis Patrick LONG, Mike Jones

Global industry is studying or deploying methods to decarbonize process heating. The electrification of process heating is one way that various industries are deploying or exploring as a method to decarbonize their processes. Shell and tube heat exchangers currently utilizing fossil fuel heated steam or hot oil is the main topic to be explored in this presentation and paper. Electrical heat exchanger replacement system examples and design will be explored. The topics explored will include materials, thermal design, mechanical design, electrical design and safety features.

Electric Heaters: Fouling, Feeders, Vaporization, PSVs, and More

Session: Electrical Heaters
Date: Tuesday, March 26, 2024
Presenter: **Dennis O. Dever, Dever Technology, LLC**
Presentation Time: 10:06 AM - 10:28 AM
Location: New Orleans Ernest N. Morial Convention Center -2024, 205

Abstract

Selected technical issues and numerical calculations associated with electric heaters are presented. Available electric heater technologies are listed. The advantages of electric heaters are described and the challenges regarding very high temperature electric heaters are summarized. The fundamental issues surrounding constant surface temperature turbulent heat transfer and constant flux turbulent heat transfer fluid will be explained with an eddy example.

Design calculations unique to electric heat applications will be presented:

- The results of a numerical study will be provided to demonstrate the problem of fouling with a radiation-based or impedance-based electric heater design.
- The results of a numerical heat transfer study for a common power feeder design will be provided.
- As an example of how direct electric heat differs from a steam heated exchanger, a comparison using vaporization will be given.
- The pressure relief calculation for a blocked pressure vessel with electric heat will be provided.

Important ohmic resistor materials will be listed with some characteristics important for their suitability in various applications. Changes and opportunities in the industry include improved radiant heat transfer software, new designs, new research, and new high temperature materials.

Introduction

There is a large increase in work and investigation into electric heaters throughout the world because of new renewable energy sources of electric power. Chemical Engineers are often not familiar with either the simple or sophisticated issues associated with electric heaters. This presentation will introduce engineers to some of these issues.

Lists and information are provided here for technologies, advantages, very high temperature challenges, resistors, radiant heater improvements, turbulent heat transfer, electric heater fouling, vaporization, and relief flow.

Electric Heater Technology List

1. Microwave. There are commercial microwave systems that heat fluids, chemicals, foods, sheet materials, and are used for drying. Systems are being promoted for the very high temperatures needed to dissociate methane.

2. Plasma. The biggest electric heater is an electric arc furnace (EAF) in Japan. In an EAF, current conducts through plasma from upper electrodes to a lower metal level. The power input is controlled, and heat radiates from the arc and ohmic heat is also generated in the metal. The electrodes attain temperatures as high as 3,000 °C (5,432 °F). Other plasma heaters ionize gas flowing past electrodes to discharge gas at higher temperatures.

3. Induction. Induction heaters create a magnetic field that generates eddy currents in a conductive material. If the process material is not conductive, a conductive material can be heated inside the equipment which will then heat the non-conductive process material by contact or radiation. This is a non-contact method used to melt metals, solder, heat silicon, and more. These can be used for very high power and temperature levels.

4. Electric Immersion Heaters. Electric immersion heaters are common electric heaters in refineries and chemical plants and are the least costly electric heaters. They resemble BEU steam heat exchangers but with sheathed U-tube elements. They heat air, water, heat transfer oil, hydrogen, and are used in reforming units.

5. Electric Radiant Heaters. Radiant heaters use high temperature resistors to radiate heat to a vessel, pipe coil, or tube coil. These are popular in high purity silicon production for the PV industry and high temperature hydrogen heaters. There are many types of radiant heaters.

6. Impedance. In impedance heaters, the live voltage in the heat source is in direct contact with the process. These work well for the very high pressures needed in supercritical water oxidizers and supercritical carbon dioxide processes. There are many types of impedance heaters.

7. Electric Conductive Fluid Heaters. In an electric conductive fluid heater, ohmic heat is generated directly in the fluid.

8. There are other technologies.

Advantages of Electric Heaters

1. High Temperatures. Electric heaters can provide very high temperatures. Whereas, even at 171.4 barg, saturated steam can only reach 353.4 C. Proven electric heater designs are available for heaters with a 3,000 °C process temperature.

2. Control. Electric heater power can be turned on and off every 0.01 second (half cycle) providing almost instantaneous power control and 0% to 100% turndown.

3. Stable Temperatures. Electric heaters do not have the fluctuating temperatures that flames have.

4. Negative Electricity Prices. There are some opportunities for very low energy prices at various locations and times because of renewable sources.

5. Regulatory Advantages. Electric heaters do not require the clean air permits that fired equipment require. Governments promote the use of electric heat to reduce carbon dioxide emissions.

Challenges with High Temperature Electric Heaters

1. Shutdowns. The electric heaters with the highest temperatures tend to use graphite electrodes or resistors and tend to be in the metals industry. In the metals industry, there are much more frequent shutdowns between runs than occur in refining or chemicals. Periodic replacement of graphite occurs, and this would be a major problem in a steady-state refining or chemical process.

2. Power Feeders. Penetrating a pressure boundary and insulation with conductors can be challenging. Electric current must conduct from the junction box where power is supplied at ambient temperatures to the interior of an electric heater, the interior of an electrically heated reactor, or the interior of other electrical equipment. Three issues must be addressed. The voltage must be isolated throughout the whole path. The ohmic heat generated in the conductor must be minimized and allowed to escape. The thermal expansion of the different materials must be managed.

3. Pressure Vessel Code. The ASME pressure vessel code limits pressure boundaries to 980 °C (1,800 °F) maximum. Only three materials are allowed for a code material at this highest temperature, and it may be impossible to purchase two of those three in a quantity less than a mill run. The boundary consists of the metal resisting the pressure of the process. This typically includes a cylindrical shell, heads, nozzles, and flanges. A pressure vessel containing a process with a temperature greater than 980 °C must have internal insulation to protect the boundary. At the highest temperatures allowed by the vessel code, there are additional challenges such as a very low allowable stress.

4. Creep. Metal gradually changes dimension under stress at high temperatures. Therefore, the pressure vessel code limits the allowable life span of equipment at very high temperatures.

5. High Temperature Material. Several types of material tend to be required within an electric heater. Examples are structural supports, insulation, current conductors, or ohmic resistors. These materials must resist the chemical environment and maintain their functional use at elevated temperatures for a long duration. The properties tend to change at high temperatures.

6. Thermal Expansion. As temperature increases different materials grow at different rates. Stress occurs if materials are constrained.

Turbulent Flow in Electric Heaters

There is a fundamental difference between constant flux heat transfer and heat transfer with a constant surface temperature. Electric heat is often considered a constant flux application. However, electric heat differs from idealized constant flux and heating with condensing steam differs from constant surface temperature. This issue will be investigated using eddy estimates in a low velocity turbulent water flow example.

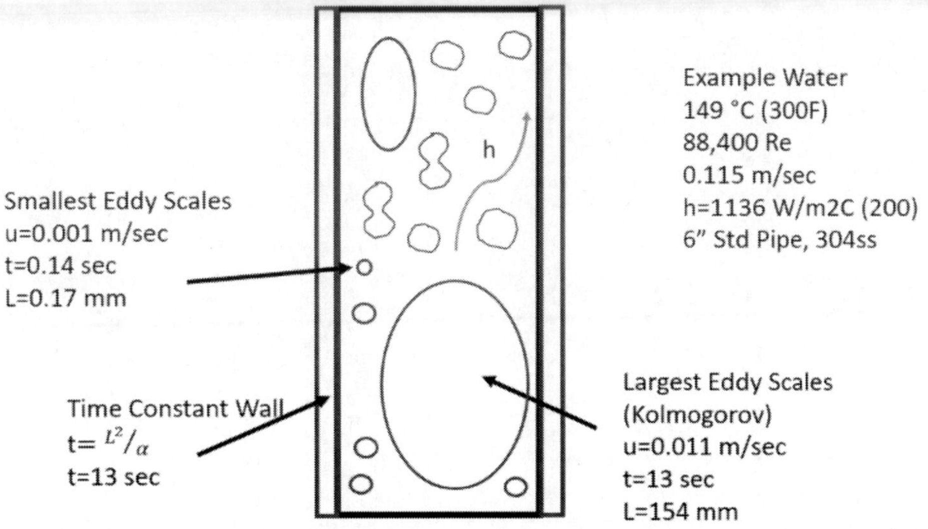

Figure 1 Eddy Size and Duration Example

In a turbulent fluid flow, there are volumes of fluid that are large enough to have enough momentum to break free of the laminar viscous forces holding them in place. These volumes are called eddies, and they are continually impinging on the wall. Local heat transfer is very high when these eddies crash into the wall. Heat transfer is low and laminar-like between eddies.

The largest and smallest eddy can be estimated using methods by the Russian mathematician Kolmogorov. This example is for water at 149 °C with an average velocity of 0.115 m/sec and Reynolds number of 88,400 in a 6-inch schedule 40 type 304 stainless steel pipe. In this low velocity example, the smallest eddy has an approximate lifespan of 0.14 seconds, and the largest lifespan is 13 seconds. This schedule 40 wall has a time constant of 13 seconds. Therefore, the thermal storage of the wall will dampen out the fluctuations of all the eddies except the largest. These three durations are order of magnitude estimates.

Because of this, the flux at the process surface of an electric heater will fluctuate somewhat because of the eddies impinging on the wall while the heat source is on the other side of the wall's thermal storage. The temperature at the process surface of a steam exchanger will fluctuate somewhat because of the eddies impinging on the wall while the condensing steam surface is on the other side of the wall. Although the idealized boundary conditions are not strictly accurate, this does not suggest that typical heat transfer correlations can be used for an electric heater. However, there are examples where the same heat transfer correlations can be used.

A typical liquid velocity in a process application will be about 0.6 m/sec to 3.6 m/sec. At these higher velocities, the largest eddies will still be about the size of the pipe inner diameter but there will be much smaller eddies. Therefore, the idealized cases are even less relevant at normal velocities.

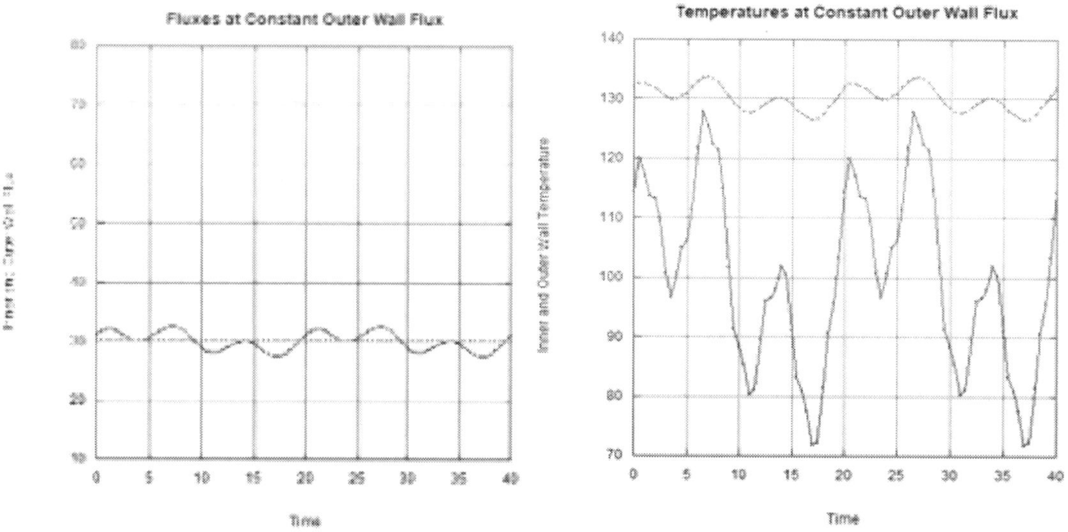

Figure 2 Flux and Temperature at Constant Flux

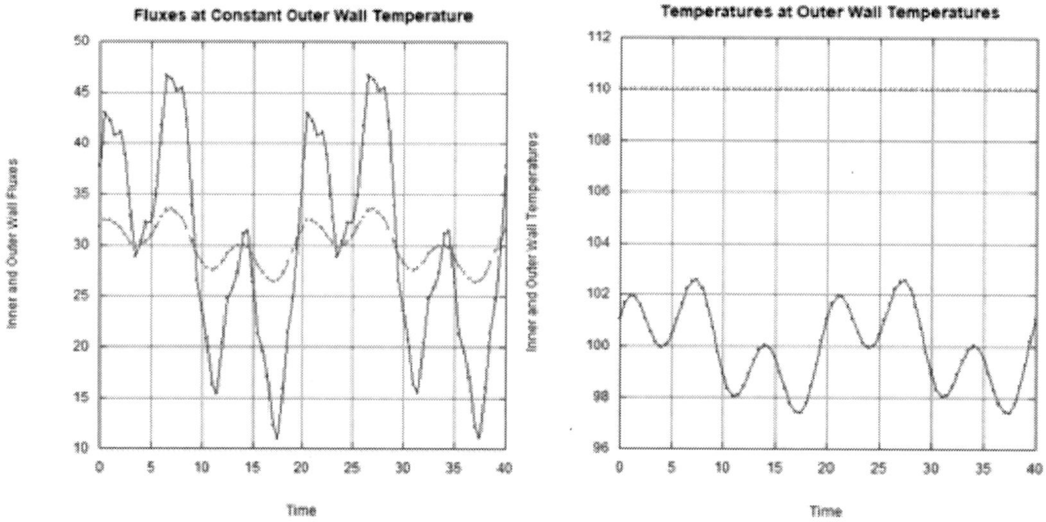

Figure 3 Flux and Temperature at Constant Temperature

Conceptual diagrams are displayed to illustrate this in Figures 2 and 3. These plots are neither data nor numerical calculations.

Numerical Study of Fouling in an Electric Heater

Fouling in a traditional heat exchanger will reduce the heat transfer flux at the location of the fouling. The total duty may be reduced so that the process requirements are threatened. In an electric heater, heat is generated in the ohmic resistors and the flux at the resistor is relatively constant. Therefore, if a thermal barrier like fouling occurs close to the electric heat source the temperature at that point will rise significantly. Therefore, fouling is a bigger problem in an electric heater and can destroy the equipment.

The effect of fouling will be investigated here using a numerical calculation and 2D geometry. One half of the 2D geometry is displayed with the other half above the horizontal line of symmetry across the top.

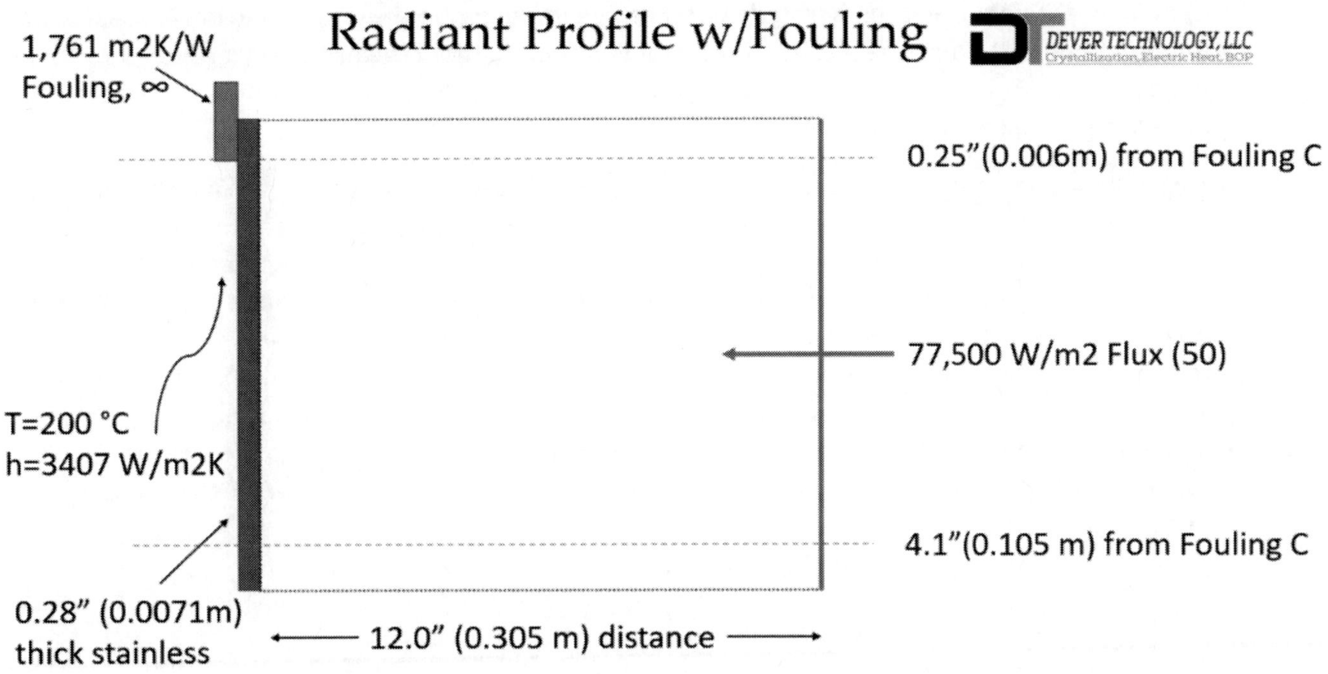

Figure 4 Inputs to the Radiant Heat Transfer Calculation with Fouling

The conditions and inputs for this numerical study of fouling with an electric heater are:

- A process fluid at 200 °C being heated by a vertical stainless-steel plate with a heat transfer film coefficient of 3,407 W/m^2K.
- The flat vertical stainless-steel plate has an emissivity of 0.60 and is 7.1 mm thick.
- A separation of 0.305 m between the hot side of the process plate and a radiant wall
- The radiant wall has an emissivity of 0.74 and an evenly distributed high flux of 77,500 W/m^2.
- At the line of symmetry, there is 12.7 mm of fouling blocking heat flux. Half of that height or 6.35 mm extends vertically into the visible symmetric plane.

This calculation is nonlinear. There is a domain with many surfaces exchanging heat by radiation and every radiant surface affects every other radiant surface. There is a separate domain internal to the process plate with many elements conducting heat, but only to the elements in direct contact. By guessing the temperatures along the line between the two domains, the two domains can be evaluated separately. Linear matrix math can calculate each domain separately and without iteration if the assumed temperatures are correct. A heat balance on the elements with assumed temperatures will indicate a net zero heat input rate if the temperature assumptions are correct. If more heat enters the elements by radiation, the assumed temperatures need to be increased for the next iteration. If more heat exits the elements by conduction, the assumed temperatures need to be reduced. The finite difference method was used for conduction and multi-surface gray body radiation was used for the surfaces with radiation.

This calculation confirms that a high source temperature is required to generate a high heat flux. The radiant surface here is about 1,100 °C. Because of the fouling, the process temperature here is 48.4 °C hotter at the fouling centerline. The fouling does not have much effect farther than 30 mm from the centerline of the fouling. The driving force for radiation is the difference between the fourth power of the absolute temperatures. At these temperatures, the fourth power of the source temperature is about 48 times greater than the fourth power of the target temperature (1104.8 °C versus 250.8 °C is 3.61×10^{12} K^4 versus 7.54×10^{10} K^4). A change in the target surface temperature has little effect. A 19% increase in the absolute target temperature would change the relative driving force from (48-1) to (48-2). These results may motivate consideration in other electric heater designs.

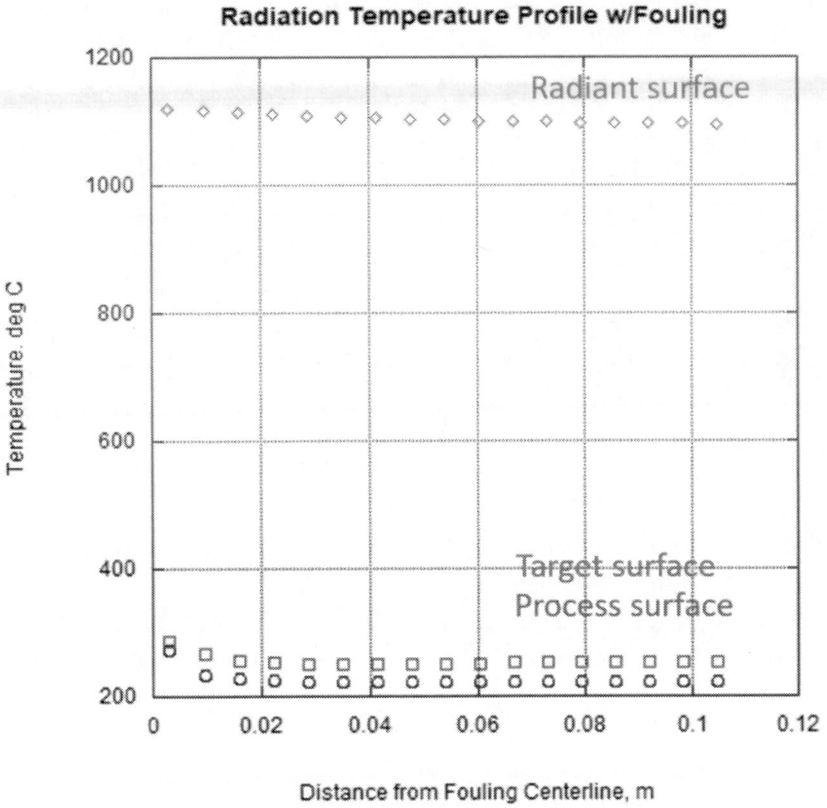

Figure 5 Temperatures with Radiation and Fouling

The same calculation was also executed without radiation as in an impedance heater. The full 77,500 w/m^2 flux was imposed directly on the hot face of the 7.1 mm process plate. In this case, the heat conduction cannot penetrate the semi-infinite fouling and can only bypass it by conduction. Again, over 30 mm from the centerline of the fouling, the fouling does not have an effect. The process temperature here is 53.3 °C hotter at the fouling centerline. This is 10% worse than the radiation case. Conduction is very effective in by-passing the fouling because the wall is thick at 7.1 mm. The hope was that radiation would bypass the fouling because of the temperature rise, but little of this happened because of the 48 to 1 driving force ratio.

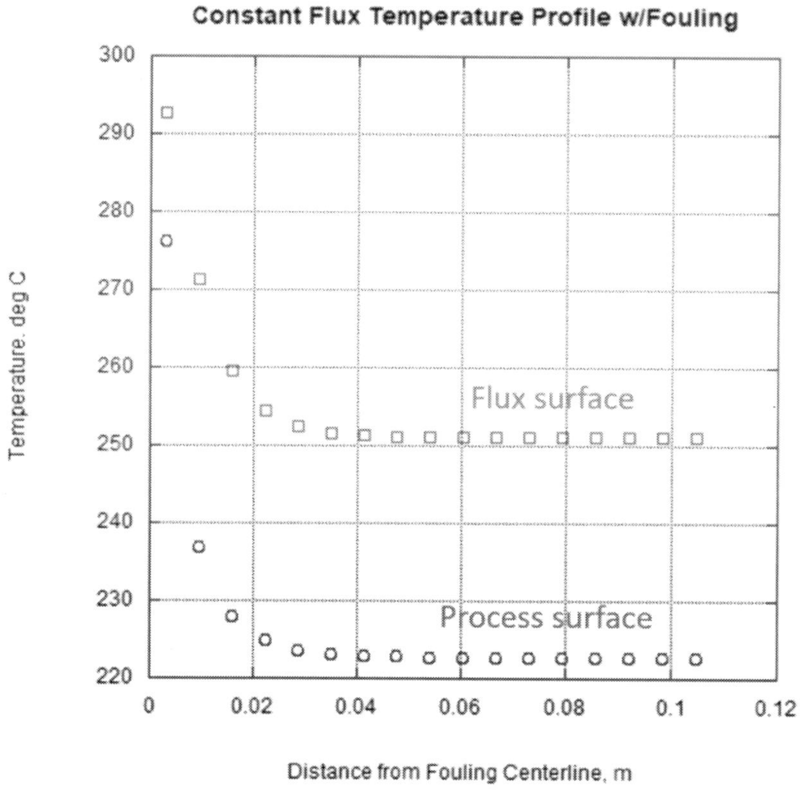

Figure 6 Temperatures with Direct Heat Flux and Fouling

Electric Heater Power Feeder and Cold Pin

The inner temperature profile for a power feed conductor (cold pin) was estimated by numerical calculation. The cold pin is the conductor transmitting current from the junction box into the interior of an electric immersion heater where the hot length starts. The agencies approving electric heaters for installation in hazardous areas require this calculation and others. The safety concern relates to high surface temperatures igniting division 2 or zone 2 hazardous atmospheres. The conduction of current through the cold pin generates heat which must be conducted through packed magnesium oxide and an outer metallic sheath and then must be transmitted by convection to the surrounding air or gas. The thermal conductivity of solid crystal Magnesium Oxide is relatively high. Loose powder has a low effective thermal conductivity. In the past, the thermal conductivity of packed and compressed Magnesium Oxide was published by major manufacturers. There is also a thermal gap resistance because the different materials expand.

This calculation demonstrated that the heat transfer that cools the sheathed element in the cold region is dominated by external free convection. The temperature across the radius of an element is approximately constant. This type of power feed can be used in other designs.

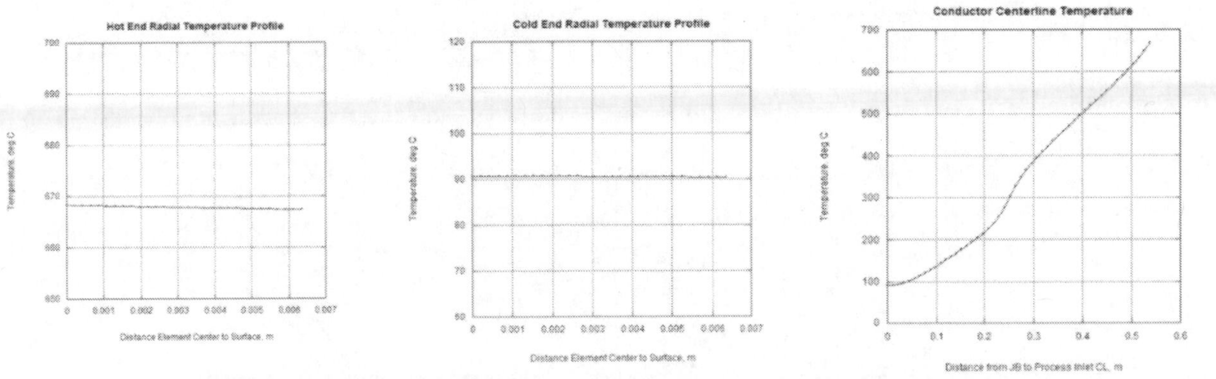

Figure 7 Temperature Profile at Cold and Hot Ends and Along Conductor

Comparison of a Steam Heated Vaporizer versus an Electrically Heated Vaporizer

The difference between process electric heaters and common heat exchangers is illustrated well by comparing vaporizers. The boiling curve for water is shaped similarly for mixtures and other liquids although the temperatures and heat fluxes are different. The x-axis is the temperature difference between the hot surface and the bulk fluid. The y-axis is the heat flux or rate. At lower temperatures, as the hot surface increases in temperature the heat flux increases greatly until the critical heat flux is achieved. For pure water, this occurs when the surface is about 30 °C hotter than the boiling fluid. At higher surface temperatures, a gas film forms at the surface and the heat transfer rate is reduced.

To design a vaporizer with steam or a constant surface temperature, conceptually draw a vertical line on the boiling curve, predict the heat flux, and select a surface and geometry that will supply the duty you need. The primary focus is to supply enough heat. These vaporizers operate in the downslope region. Notice that higher temperatures tend to reduce capacity which has resulted in many bad initial control loops.

To design a vaporizer with electric heat, conceptually draw a horizontal line and select a geometry that will not overheat the process or hardware at any point. The focus with electric heat is to avoid overheating. These vaporizers cannot operate in the downslope region. If the design ever allows the flux to exceed the maximum, the heater will be destroyed. Above the critical heat flux, the heat cannot all be removed, and it will accumulate and overheat before it can reach the radiant region.

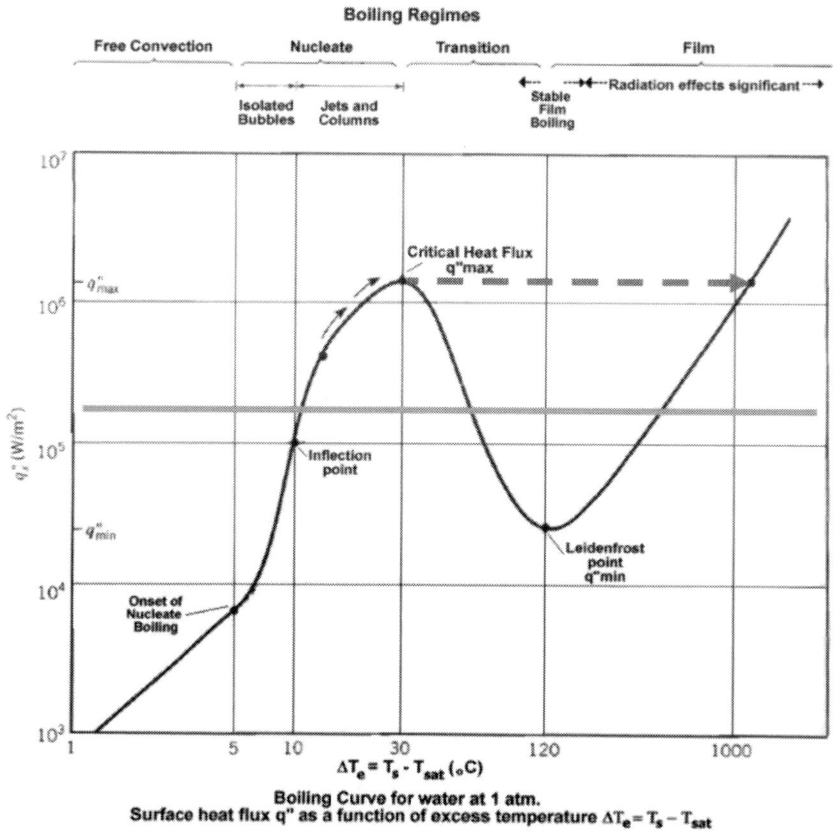

Figure 8 Constant Flux/Electric Vaporizer Example (Wikipedia – Venny85)

Electric Heater Pressure Relief with Thermal Expansion

A Mollier diagram for water is used to illustrate single-phase blocked fluid electric heater pressure relief. This concept applies to a liquid, a gas, a supercritical fluid, or a two-phase fluid that is slowly heated into the supercritical region. The X-axis is enthalpy, and the Y-axis is pressure. Along the liquid region on the left are constant density lines where a very small heat input quickly raises the pressure.

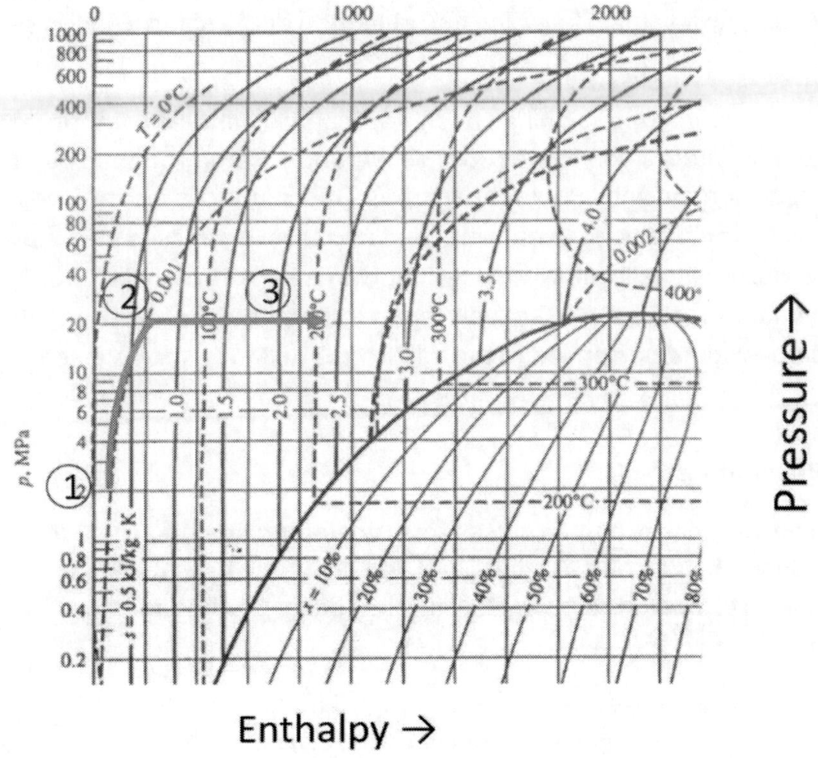

Figure 9 Blocked Volume Electric Heater Relief

If a fluid is trapped in a fixed volume electric heater, then the fluid density equals the trapped mass divided by the vessel volume. If the distortion of the vessel is negligible, the relief scenario involves movement from point 1 to point 2 along the constant density line. It is conservative to neglect the distortion of the vessel but alternately the volume distortion can be estimated. Point 1 is the normal operating point. Point 2 is at the safety valve pressure set point where the valve opens. The heat addition or power from point 1 to point 3 is based on a simple ohmic calculation. For a steam heated exchanger, the heating from point 1 to 3 would be an irregular free convection and dynamic heat transfer mechanism with a changing driving force. In this case, relief flow equals the heat input rate times the ratio of the incremental mass change per incremental heat addition. Vessel volume is not required in the final solution. A constant pressure after the PSV opens is used here. A dynamic flow calculation could be considered to estimate the slope from point 2 to 3. Generally, a Mollier diagram is not available for a process mixture so property correlations can be used to create a 3-point or 5-point Mollier diagram near points 1, 2, and 3. In this equation, $\dot{m}$ equals mass flow (kg/hr), Q equals heat input (kJ/hr), β equals thermal expansivity (1/C), and C_p equals heat capacity (kJ/kgC).

$$\dot{m} = \frac{Q\beta}{C_p}$$

This method provides a flow value. The sizing and selection of the safety valve is separate. Consider the following,

- Single phase thermal expansion has been traditionally handled without calculations by choosing the smallest safety valve (PSV) with a Code Stamp. Trapped electric heat is like thermal expansion but with greater heat input. This method is an improvement.
- Large safety factors are used to convert yield and tensile stresses into allowable stresses.
- Vessel internal pressure is allowed to drift upwards by 10% of the set point during relief flow.
- Metal allowable stresses are effectively larger in a highly dynamic scenario.
- Extreme slopes could be applied to the line from 2 to 3 to investigate sensitivity.

Electric Heater Resistor List

Nickel Chromium alloys are very common resistors for electric heaters. The 70/30 alloy is suitable for higher temperatures than the 80/20 alloy but it should not be used in Magnesium Oxide. The resistance change with temperature is very low for the Nickel Chromium alloys which is a significant benefit.

The maximum temperature allowed on a pressure vessel material is 980 °C which is available with Alloy 800H. Two other code materials are allowed at 980 °C but commercial availability for them is very challenging. There are serious challenges such as creep and very low allowable stress at temperatures near 980 °C for all code materials. This issue is important for type 1 impedance heaters because the pressure boundary heats the fluid.

Tungsten and APM are suitable at higher temperatures, but tungsten quickly oxidizes in air. Tungsten has a 550% change in resistance with temperature which requires special power controls.

Silicon Carbide and Molybdenum Disilicide may be used at very high temperatures, but the resistivity curves require extra cost and effort. Some chemical environments are restricted.

Designs at process temperatures of 2,000 °C to 3,000 °C tend to be in the metals industry and use graphite. The graphite is used for electrodes, resistors, and insulation. The graphite is frequently replaced which is unacceptable in industries with long periods between shutdowns.

Electric Radiant Heater Improvements

A helical pipe or tube coil heated by hot ohmic resistors near the wall has been a common radiant design for several decades. There are four big improvements versus the old designs.

- The helical geometry with outer elements will result in hot spots at high heat fluxes. At high temperatures or high fluxes, there are better geometries.
- Improvements in ohmic resistor materials
- Excellent radiation and view factor software
- Good research and correlations for helical flow pressure drop and heat transfer

Published view factors are not available for the most common commercial geometries like helices or radial patterns of cylinders in an outer cylindrical shell. In one case, the element-to-element view factor for a helix design was 0.22 whereas a manual estimate was 0.05. A high view factor from radiant elements to themselves is undesirable.

Flux based design methods by experienced engineers are good for estimates, preliminary proposals, and initial iterations. However, avoid flux-based designs for high temperatures.

Conclusions

At typical process temperatures an electric radiant heater will probably not greatly improve the peak temperatures that can occur with localized fouling in a more direct electric heater design. At very high target surface temperatures, radiation may provide benefits. The extent to which wall conduction can allow heat to bypass fouling was demonstrated with one example.

A numerical calculation indicated that external free convection dominates the cooling needed in the power feeder to an immersion heater. The need to design and control an electrically heated vaporizer below the critical heat flux was described. A safety valve sizing method was provided for an electric heater with a blocked fluid. Four improvements in modern radiant heaters versus legacy designs were listed.

Dennis O. Dever, PE, PhD

ddever@devertechnology.com

www.devertechnology.com

48a Dr. Ken Bell - the Person, the Expert, the Educator and the Mentor. How His Ideas Are Pertinent in Today's Challenges and New Frontiers

Anil Gokhale Sr.

This presentation in the Dr. Ken Bell memorial session will begin and focus on author's interaction with him through years at Oklahoma State university in the graduate programs. Several specific instances highlighting Dr. Bell's passion for his expertise, and desire to pass on the knowledge will be presented. There will be examples of his far-reaching thinking about AI and alternate energy sources that are applicable even today, several decades later.

It will touch on some of the unique heat transfer challenges faced by the new technologies such as RAPID manufacturing, the energy transition initiatives and decarbonization efforts - covering conduction, convection and radiation.

48b Design and Optimization of High Temperature Heat Exchangers

Blazo Ljubicic

Abstract

This paper summarizes issues concerning selection, design, construction, operation, and maintenance of High Temperature Heat Exchangers (HTHE).

Historically, HTHEs were mainly used for gas turbine engine recuperation and flue gas recuperation. Recent application areas are found in low emission power plants, hydrogen production, high temperature fuel cell systems, and Carbon Dioxide (CO_2) capture and sequestration (CCS).

While it is evident that weight and volume are not vitally important in stationary power plants, there is a strong incentive to reduce both weight and volume of HJTHE because considerations of shielding, remote handling, fluids inventory, hazards, control response rates, and costs. For marine power generation weight and volume become critically important.

Depending on the HE construction, there is a threshold for the maximum operating temperature. Beyond this temperature level, the HE cannot be used, or special design and material considerations are required. This threshold is usually considered to be 600 °C.

This paper presents information about issues of high-temperature Shell-and-Tube heat exchangers, for services up to 1100°C. Materials selection, confidence in heat transfer and pressure drop predictions, mechanical issues associated with thermal strain cycling caused by changes from low to high load were discussed. The current gaps in design manufacturing, and operation in this field were identified with the objective to improve high-temperature heat exchanger efficiency, economic feasibility, and safety.

Extended Abstract for

The Future of Shellside Thermal Performance Research

Thomas Lestina, Heat Transfer Research, Inc.

3rd Topical Conference on Heat Exchangers
Special Session in Honor of Professor Kenneth Bell
Monday, March 25, 2024: 3:30 PM - 5:00 PM

Process industries rely heavily on shell-and-tube heat exchangers. The widely accepted standards of the Tubular Exchanger Manufacturer's Association [1] provide fabrication guidelines which can be met by a wide range of manufacturers, many with limited sophisticated machinery or specialized skills. The ease of design and fabrication along with its widespread acceptance are the key advantages of this established technology. Originally, thermal designs for the shell side were performed by calculating an overall heat transfer coefficient and pressure drop for an ideal crossflow bundle and de-rating the result by applying an "experience factor." This practice did not instill confidence in the thermal design, and research into a more fundamental approach was undertaken.

Shellside flow is complicated and systematic research over an extended time was needed to develop commercially viable predictive methods. The first extensive research study of shellside thermal performance was performed at the University of Delaware from 1947 to 1963. The project was quite comprehensive involving 34 researchers and producing 39 publications focusing on non-ideal flow patterns caused by the baffles [2]. From this work, Bell [3] first presented a design method for single-segmental baffled exchangers. Over the years, the method has been modified, with the most commonly used version documented in the *Heat Exchanger Design Handbook* [4]. The method consists of calculating the heat transfer coefficient and pressure drop for an ideal bundle based on the gap velocity (maximum mass flow rate in the gap between the tubes) and correcting these values based on functions which account for non-ideal construction practices and process conditions:

$$h_s = h_{ideal} J_c J_l J_b J_s J_r \qquad (1)$$

$$\Delta P_s = \Delta P_{bi}(N_b - 1)R_l R_b + \Delta P_{wi} N_b R_l + 2\Delta P_{bi} \frac{N_c + N_{cw}}{N_c} R_b R_s \qquad (2)$$

Shellside research continued at Heat Transfer Research, Inc. (HTRI), a consortium of companies dedicated to the pursuit of heat transfer research and commensurate software development activities. Palen and Taborek [5] provide the results of this early HTRI research which shows that the HTRI method is more accurate than available open literature methods. At that time, the standard for predictive accuracy was ±30%, implying that if a predicted heat transfer coefficient or friction factor was within 30% of the actual value, it was sufficiently accurate for design. As a result, thermal design margins in excess of 30% were typical. HTRI shellside research has continued since this early publication, and the method has been modified to include different baffle types (such as double segmental and no-tube-in-window) and to improve the predictive accuracy for selected applications.

Commercial heat exchanger thermal performance software is based on stream analysis which is an iterative shellside calculation procedure where the ideal flow stream through the bundle (the B-stream) is balanced with bypass flow streams [6]. As illustrated in Figure 1, the bypass flow streams consist of the A-stream (in the gap between the tube and baffle hole), the C-stream (outside the tube bundle), the E-stream (in the gap between the baffles and the shell), and the F-stream (in the passlane).

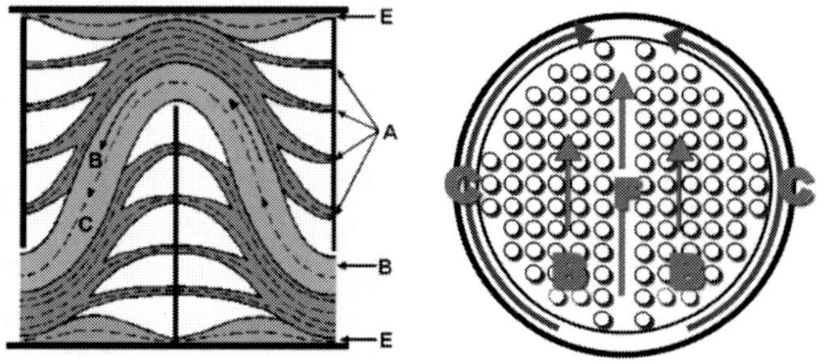

FIGURE 1. SHELLSIDE FLOW STREAMS USED IN STREAM ANALYSIS

Research continues for shellside heat transfer and pressure drop to improve accuracy and develop performance improvements. The following open areas summarize broad areas of ongoing research:

Two-phase flow. The stream analysis was developed with single-phase vapors and liquids. Experimental studies and numerical analyses using computation fluid dynamics (CFD) have been undertaken to address shellside two phase flow. Of particular importance is classifying flow patterns. Grant and Chisholm [7] published a flow regime map based on early experimental studies. Research continues in the area to more accurately predict heat transfer, pressure drop and potential instabilities.

Stream mixing. Bypass streams have a different temperature profile than the B-stream. The impact of these different temperature profiles is illustrated in Figure 2.

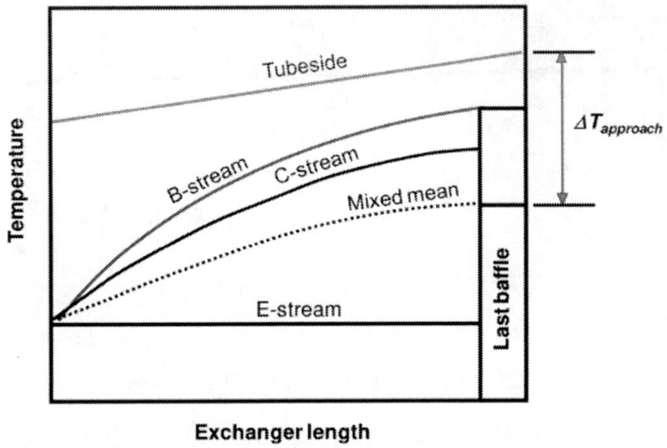

FIGURE 2. DIFFERENT SHELLSIDE TEMPERATURE PROFILES FOR FLOW STREAMS

The temperature difference between the ineffective flow streams and the tube wall is significantly greater than the temperature difference between the B-stream and the tube wall. The mean temperature difference for thermal performance calculation should be based on the B-stream temperature difference rather than on the mixed mean shellside temperature. However, typical design software considers only the mixed mean temperature profile from shell inlet to outlet, using correction functions to handle the impact of the different temperature profiles for the ineffective flow streams on the mean temperature difference. These correction functions can have a large impact on thermal performance predictions for high effectiveness designs. High effectiveness designs generally have low approach temperatures and large shellside temperature changes. CFD is a useful tool to better estimate the mixing and resulting performance of these high effectiveness designs.

Non-TEMA baffle arrangements. TEMA requirements are based on segmental baffle designs. Research has shown that segmental baffle designs can produce regions of stagnant flow and recirculation. Alternate baffle arrangements can improve flow distribution and thermal performance. Examples of some of these non-segmental baffles are shown in Figure 3.

Each of these non-TEMA baffle technologies uses a longitudinal flow component (along the shell axis) to reduce pressure drop, increase velocities, and eliminate stagnant zones. To be considered successful, a design should minimize bypass flow. So that the tube bundle can be as full as possible, annular distributors are often used to distribute flow in the inlet and outlet regions without any tube removal under the nozzle. In addition, blanking strips or baffle rings are used to minimize the gap at the outer edge of the bundle. In some cases, the bundle is shrouded to effectively eliminate any longitudinal bypass flow. Stream analysis methods are still under development for these alternate baffle technologies, and designs are often handled with proprietary methods. As the non-TEMA baffle types become more popular, the need for fundamental thermal analysis methods increases.

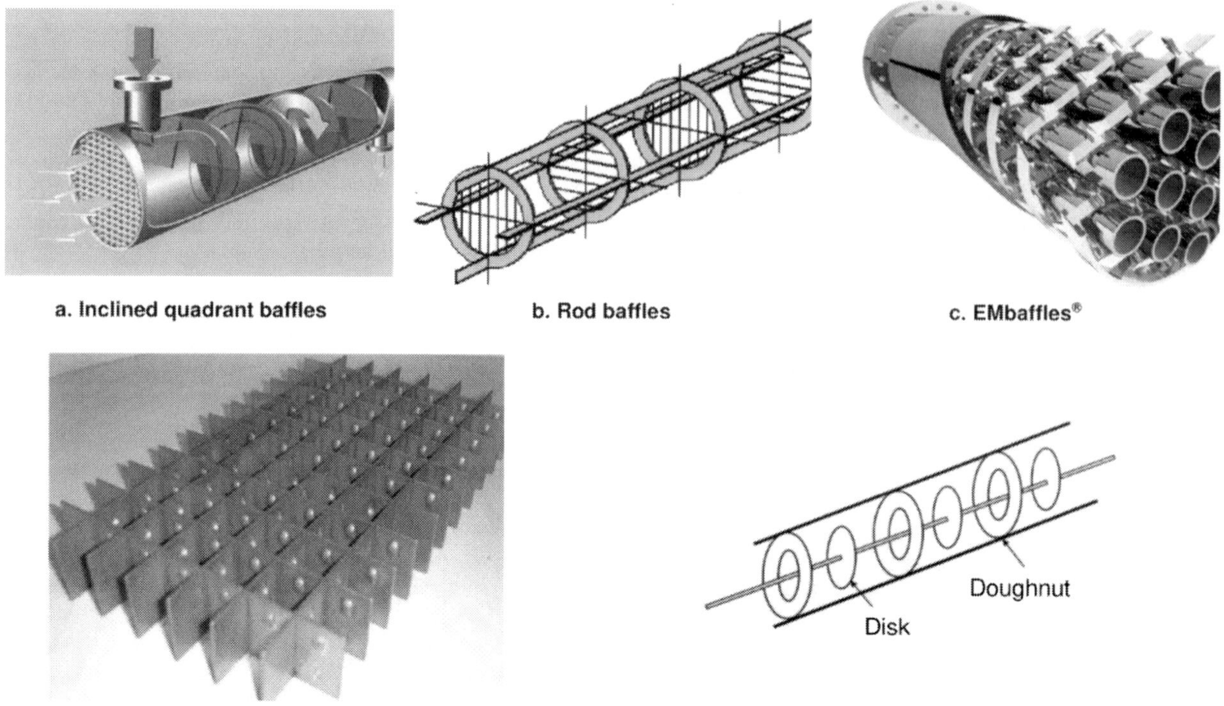

a. Inclined quadrant baffles b. Rod baffles c. EMbaffles®

d. Strip baffles e. Disk-and-doughnut baffles

FIGURE 3. NON-TEMA BAFFLE TYPES

Enhanced tubes. The use of tubes which enhance or increase the shellside heat transfer coefficient has been increasing. These enhanced tubes include TWISTED TUBE®[1], finned tubes, double-cut tube surfaces, corrugated tubes, porous metallic coatings, and other proprietary surface texture techniques. These enhanced surfaces are often selected for applications in which the temperature difference between the hot and cold fluids is low (less than 5 °C) and the design includes little excess margin. When these technologies are used, manufacturing clearances lower than TEMA's are often needed to ensure sufficient confidence in the thermal performance of the final design. Stream analysis methods may need to be modified for these enhanced tube technologies when tight manufacturing clearances are applied.

Fouling mitigation. Shellside designs which reduce the potential for fouling improve process efficiency and increase throughput. By reducing flow stagnation, each of the non-TEMA baffle types shown in Figure 3 reduce fouling potential. Other geometries are a subject of ongoing research. Additive manufacturing and self-cleaning configurations are areas of further study.

Trends in process design demand higher accuracy because lower temperature differences between the hot and cold fluids are specified and higher efficiencies are required. Ongoing shellside research is needed to meet these demands. Baffle designs which promote efficient countercurrent flow and improved manufacturing practices should facilitate competitive thermal designs in the future.

Nomenclature

f, Fanning friction factor
h_{ideal}, Heat transfer coefficient for ideal bundle, W/m^2 K
h_s, Shellside heat transfer coefficient, W/m^2 K
j, Colburn j-factor
J_b, Correction factor for bundle bypass effects of C- and F-streams
J_c, Correction factor for baffle cut and spacing

[1] TWISTED TUBE is a registered trademark of Koch Heat Transfer Company, L.P.

J_l, Correction factor for leakage effects
J_r, Correction factor for laminar flow
J_s, Correction factor for variable baffle spacing at inlet and outlet
N_b, Number of baffles
N_c, Number of tuberows crossed in one crosspass
N_{cw}, Number of tuberows in window,
R_b, Correction factor for bypass flow
R_l, Correction factor for baffle leakage effects (A- and E-streams)
R_s, Correction factor for entrance and exit section
Re_s, Shellside Reynolds number
V, Fluid velocity, m/s

Greek
ΔP_{bi}, Ideal crossflow bundle pressure drop, Pa
ΔP_s, Pressure drop in shell excluding nozzles, Pa
ΔP_{wi}, Ideal window flow pressure drop, Pa
$\Delta T_{approach}$, Approach temperature which is the minimum temperature difference between hot and cold streams, °C
ρ, Fluid density, kg/m^3

References

1. Tubular Exchanger Manufacturers Association, *Standards*, 10th ed., Tarrytown, New York, 2019.
2. Bell, K., Final Report of the Cooperative Research Program on Shell and Tube Exchangers, Bulletin 5, University of Delaware, June 1963.
3. K. J. Bell, Exchanger design—based on the Delaware Research Program, *Petro/Chem Engineer*, 26 – 40, October 1960.
4. G. F. Hewitt, ed., *Heat Exchanger Design Handbook*, Begell House, New York, 1998.
5. J. W. Palen and J. Taborek, Solution of shellside flow pressure drop and heat transfer, *CEP Symp. Series* **92**(66), 53 – 63, 1969.
6. T. Tinker, Shellside heat transfer characteristics of segmentally baffled shell and tube heat exchangers, ASME Paper No. 47-A-130, New York, 1947.
7. I. D. R. Grant and D. Chisholm, Two phase flow on the shell side of segmentally baffled shell-and-tube exchangers, *ASME J. of Heat Transfer* **101**, 38-42, 1979.

48d Process Heat Transfer for Enhanced Energy Efficiency and Productivity: Past, Present and Future.

Chandrakant (CB) Panchal

The process heat transfer has come long way from 50-years ago to the present design and development status. Prof. Bell's work was a crucial part of the design and development of process heat transfer and provides foundation for achieving decarbonization goal of the process industry. Dr. Jerry Jaborek, first HTRI director, in his 1979 D.Q. Kern Award lecture provided an overview of evaluation of heat exchanger design technique. He indicated that more shell-and-tube heat exchangers have been designed in that time using methods described in *Kern's Process Heta Transfer*. The customary sources of new developments for practical applications disappeared and confidential tools such as HTRI, HTFS and FPRI became the design tools. The future trends would consist of: a) sophisticated design methods; b) reliable physical and thermodynamic properties; c) design program logic based on process configuration; d) shop-ready heat exchanger drawings and e) engineering judgement decision. The present status of process heat transfer consists of a) alternate heat exchangers, plate-frame and plate-fin; b) reliable physical properties by linking with property databases; c) interfacing ASPEN Plus process simulation; d) sensitivity analysis to evaluate the impact on process conditions and e) economic analysis. Prof. Bell's work on phase-change heat transfer, specifically multi component condensation, is well recognized. His work on process heat transfer expanded to energy conversion systems. Selected selection of publications of this author with Prof. Bell elucidate: 1) condensation and flow boiling in spirally fluted tubes; 2) simultaneous production of desalinated water and power with hybrid-OTEC system using aluminum plate-fin condenser; 3) Nusselt type condensation on triangular fluted tubes; 4) plate-fin condensation of low-pressure steam in the presence of non-condensable gases; 5) spirally fluted tube heat exchanger as a condenser and evaporator and 6) condensation keynote presentation at *International Heat Transfer Conference*, 1978. Prof. Bell had supervised a good number of Master and PhD students and published in the open literature. Prof. Bell's research provides foundation for future process heat transfer consisting of: a) process intensification by integrating reaction and heat transfer, specifically in reactive distillation column; b) design of reboilers to minimize maldistribution of liquid and vapor phases causing localized fouling; c) heat integrated distillation for enhanced energy efficiency; d) heat integration in capture of carbon dioxide and conversion to value-added products. In summary, this presentation links process heat transfer from past to present leading to future with the focus on decarbonization in the process industry.

194a Better Cleaning Is Less Cleaning; Thermal Cleaning of Heat Exchangers Explained

Robert Mol, Philippe Vandewalle, Appie Peters

Better cleaning is less cleaning; thermal cleaning of heat exchangers explained

The Thermo-Clean group has developed an improved way of removing fouling from heat exchangers using thermal cleaning. This resulted in a safe and very effective cleaning technology with astonishing results in operational lifespan and energy savings.

By the use of controlled heat, a very thorough cleaning takes place, even in places where the typical high pressure cleaning has no access to. After thermal cleaning, heat exchangers will practically come up to their original specs. The result: perfect heat transfer, longer use in production and no residue mixing with your products.

This cleaning method is so effective, because the combination of the low oxygen level and the perfect temperature control ensures that the organic component of the fouling is converted into flue gases that will be used as a source of energy within our thermal cleaning plant. 1 kilo of contamination thus turns into 50 grams of easy-to-remove dust residue.

Furthermore, thermal cleaning is fast, so short delivery times are possible during shutdowns. After all, better cleaning means less cleaning. As a result, the time between expensive shutdown cleanings can be extended.

Advantages:

- Perfect cleaning quality (up to 100%), even on the inside
- Benefit à performance levels like new
- Fixed duration of the cleaning à stable cleaning costs
- Reduction of greenhouse gas emissions
- No wastewater problems
- Less energy consumption

Thermal cleaning: makes it possible to clean everywhere!

- Inside the tubes
- Around the tubes
- Between the tubes & the shell
- Inside tubes with static mixers

In addition to heat exchanger cleaning, this method can also be perfectly used for cleaning extruder parts, filters, pipelines, tanks and furthermore all metal parts with organic or partially organic contamination.

194b Heat Exchanger Tube ID Cleaning Using Vbx Technology

Nathan Conrad

USA Debusk has pioneered a new method of heat exchanger tube ID cleaning to meet NDE testing cleanliness requirements using our VBX technology. Current common practices for exchanger tube cleaning to meet NDE standards include multiple conventional hydroblasting attempts, high pressure 40K hydroblasting, grit/abrasive blasting and offsite specialty cleaning. VBX technology offers advantages to each of these common practices saving time, reducing waste, reducing risk, and reducing cost. USA Debusk has deployed VBX heat exchanger tube cleaning on over 1 million tubes with a 100% Eddy Current and IRIS test pass rate after one cleaning attempt.

194c The Cost of Incomplete Cleaning

Roberto Tomotaki, Byron Kieser, Himanshu Joshi

The fouling of heat exchangers has a well understood impact on operating efficiency as it can increase energy consumption, emissions, maintenance costs, and above all may limit throughput or product quality. Operators expend significant effort and cost to address fouling through maintenance and turnaround cleaning. In a program of trials, it has been demonstrated that the traditional methods commonly employed often fail to clean heat exchangers back to "clean performance" levels. What is the cost of the fouling that is left behind by incomplete cleaning? The benefits of a thorough cleaning and how different cleaning techniques affect these benefits will be discussed in this presentation, along with a system for accounting for all cleaning and fouling costs.

It is often impossible to assess, on the washpad, whether a heat exchanger is completely clean, and because different cleaning techniques have varying degrees of effectiveness and cost, we frequently inadvertently settle for a partial cleaning. Before choosing a cleaning technique, it is important to take into consideration all cleaning-related expenses as well as the impact of incomplete cleaning. We will examine the economics of chemical cleaning, hydroblasting, and ultrasonic cleaning as they apply to the case of a crude preheat exchanger where fouling results in all the effects mentioned. We will also give a brief overview of the various cleaning methods, including how they are applied, how expenses are affected, and when they should be done.

AUTHOR INDEX